Detection and Data Analysis in Size Exclusion Chromatography

ACS SYMPOSIUM SERIES 352

Detection and Data Analysis in Size Exclusion Chromatography

Theodore Provder, EDITOR

The Glidden Company

Developed from a symposium sponsored
by the Division of Polymeric Materials Science and Engineering
at the 191st Meeting
of the American Chemical Society,
New York, New York,
April 13–18, 1986

American Chemical Society, Washington, DC 1987

Library of Congress Cataloging-in-Publication Data

Detection and data analysis in size exclusion .
chromatography.

(ACS symposium series; 352)

Includes bibliographies and index.

1. Gel permeation chromatography—Congresses.

I. Provder, Theodore, 1939-   . II. American
Chemical Society. Division of Polymeric Materials:
Science and Engineering. III. American Chemical
Society. Meeting (191st: 1986: New York, N.Y.)
IV. Series.

QD272.C444D47   1987   547.7'046   87-19480
ISBN 0-8412-1429-8

Sep/ae

Chem

Copyright © 1987

American Chemical Society

All Rights Reserved. The appearance of the code at the bottom of the first page of each chapter in this volume indicates the copyright owner's consent that reprographic copies of the chapter may be made for personal or internal use or for the personal or internal use of specific clients. This consent is given on the condition, however, that the copier pay the stated per copy fee through the Copyright Clearance Center, Inc., 27 Congress Street, Salem, MA 01970, for copying beyond that permitted by Sections 107 or 108 of the U.S. Copyright Law. This consent does not extend to copying or transmission by any means—graphic or electronic—for any other purpose, such as for general distribution, for advertising or promotional purposes, for creating a new collective work, for resale, or for information storage and retrieval systems. The copying fee for each chapter is indicated in the code at the bottom of the first page of the chapter.

The citation of trade names and/or names of manufacturers in this publication is not to be construed as an endorsement or as approval by ACS of the commercial products or services referenced herein; nor should the mere reference herein to any drawing, specification, chemical process, or other data be regarded as a license or as a conveyance of any right or permission, to the holder, reader, or any other person or corporation, to manufacture, reproduce, use, or sell any patented invention or copyrighted work that may in any way be related thereto. Registered names, trademarks, etc., used in this publication, even without specific indication thereof, are not to be considered unprotected by law.

PRINTED IN THE UNITED STATES OF AMERICA

_SD_
_12-14-87_
_cu_

QD272
C444
D471
1987
CHEM

ACS Symposium Series

M. Joan Comstock, *Series Editor*

1987 Advisory Board

Harvey W. Blanch
University of California—Berkeley

Alan Elzerman
Clemson University

John W. Finley
Nabisco Brands, Inc.

Marye Anne Fox
The University of Texas—Austin

Martin L. Gorbaty
Exxon Research and Engineering Co.

Roland F. Hirsch
U.S. Department of Energy

G. Wayne Ivie
USDA, Agricultural Research Service

Rudolph J. Marcus
Consultant, Computers &
    Chemistry Research

Vincent D. McGinniss
Battelle Columbus Laboratories

W. H. Norton
J. T. Baker Chemical Company

James C. Randall
Exxon Chemical Company

E. Reichmanis
AT&T Bell Laboratories

C. M. Roland
U.S. Naval Research Laboratory

W. D. Shults
Oak Ridge National Laboratory

Geoffrey K. Smith
Rohm & Haas Co.

Douglas B. Walters
National Institute of
    Environmental Health

Foreword

The ACS SYMPOSIUM SERIES was founded in 1974 to provide a medium for publishing symposia quickly in book form. The format of the Series parallels that of the continuing ADVANCES IN CHEMISTRY SERIES except that, in order to save time, the papers are not typeset but are reproduced as they are submitted by the authors in camera-ready form. Papers are reviewed under the supervision of the Editors with the assistance of the Series Advisory Board and are selected to maintain the integrity of the symposia; however, verbatim reproductions of previously published papers are not accepted. Both reviews and reports of research are acceptable, because symposia may embrace both types of presentation.

Contents

Preface

The FIELD OF SIZE EXCLUSION CHROMATOGRAPHY (SEC) remains a viable and lively area of polymer characterization. Over the past several years, there has been considerable research activity in the area of SEC detection and data analysis in order to obtain more comprehensive information concerning the composition and molecular architecture of complex polymer systems.

In part, this has been brought about by the efforts to produce unique polymeric materials from a constrained set of commercially available building blocks. These constraints resulted from government legislation in the areas of clean air, toxic substances, hazardous wastes, etc. As a consequence of being restricted to a narrow set of building blocks, it becomes critical to understand how a polymeric material is put together (composition, structure, molecular architecture, morphology) in order to relate fundamental properties to a polymeric material's performance. Therefore, the use of concurrent detectors in SEC along with sophisticated data analysis methods to unravel the nature of complex polymers is growing. Advances in electronics and computer technology are catalyzing the activities of detector development and data analysis.

The detection and data analysis activities in the field of SEC applied to polymeric materials is expected to grow in the future. Improved detectors and data analysis systems will become commercially more available as a result of the current research activities in selected industrial and academic labs.

I thank the authors for their effective oral and written communications, and the reviewers for their critiques and constructive comments.

THEODORE PROVDER
The Glidden Company
Strongsville, OH 44136

ix

GENERAL CONSIDERATIONS

Chapter 1

An Overview
of Size Exclusion Chromatography
for Polymers and Coatings

Cheng-Yih Kuo and Theodore Provder

The Glidden Company, Research Center, 16651 Sprague Road,
Strongsville, OH 44136

Recent technological advances have sparked a new
level of activity in the field of Size Exclusion
Chromatography (SEC). These include: 1) high
performance/high speed column technology, 2) the
development and increased use of simultaneous
multiple in-line detectors such as differential
refractometer, ultraviolet and infrared
spectrophotometric detectors, viscometers, low angle
laser light scattering, and mass detection, and 3)
the application of minicomputer and microcomputer
technology for instrument control and data analysis.
These developments in turn have led to new improved
applications of SEC as well as higher quality
information. In this paper, the SEC separation
mechanism, molecular weight calibration methods
including the use of hydrodynamic volume, instrument
spreading corrections and polymer chain branching
calculations will be discussed. Quantitative and
qualitative examples of the application of multiple
detectors will be given. Finally, there will be
some discusison of the requirements necessary for
high resolution SEC analysis of oligomers and
examples will be shown.

Polymer chemists and coatings formulators are continually being
called upon to tailor-make coatings systems which require
polymers having specifically designed molecular architectures and
physical properties. Knowledge of the molecular weight and
molecular weight distribution (MWD) of the polymer components in
a coatings system is essential for the optimization of polymer
design for specific end-use properties. Since its introduction
over two decades ago,(1) gel permeation chromatography (GPC) or
size exclusion chromatography (SEC) has become an important and
practical tool for the determination of the MWD of polymers. A
large number of studies has been published on the use of SEC in
plastics, elastomeric and coatings systems. With the advent of
high efficiency columns, the resolution in the lower molecular

0097–6156/87/0352–0002$07.75/0
© 1987 American Chemical Society

weight region (molecular weights in the range of 200 to 10,000) has been greatly improved and the speed of analysis increased. These features make high performance SEC (HPSEC) an indispensable characterization tool for the analysis of oligomers and polymers in environmentally acceptable coatings systems.

SEC Separation Mechanism

Size exclusion chromatography is a liquid chromatography method, whereby, the polymer molecules are separated by their molecular size or "hydrodynamic volume" in solution as solvent elutes through a column(s) packed with a porous support. The degree of retention of the polymer molecules in the pores is the phenomenon which affects the separation. Smaller molecules are retained in the pores to a greater degree than the larger molecules. As a result the largest size molecule (or the molecule having the greatest hydrodynamic volume) elutes from the column first followed by the smaller molecules. The volume of liquid at which a solute elutes from a column or the volume of liquid corresponding to the retention of a solute on a column is known as the retention volume (V_R) and can be related to the physical parameters of the column as follows:

$$V_R = V_o + KV_i \qquad (1)$$

where V_R = retention volume of the solute
V_o = interstitial volume (dead volume) of the column
V_i = internal solvent volume in the pores
K = the distribution coefficient, based upon the relative concentrations between phases.

The total column volume V_T is given by

$$V_T = V_o + V_i \qquad (2)$$

Therefore, the retention volume is expressible in terms of the two measurable quantities V_o and V_T as

$$V_R = V_o(1-K) + KV_T \qquad 0 \leq K \leq 1 \qquad (3)$$

The dependence of molecular size in solution upon retention volume is schematically illustrated in Fig. 1. The void volume V_o corresponds to the total exclusion of solute molecules from the pores. The excluded solute molecules are significantly larger than the largest available size pores. Between V_o and V_T the solute molecules are selectively separated based on their molecular size in solution. Beyond the total column volume V_T, separation will not be achieved by a liquid exclusion chromatography mechanism. If molecules appear to separate beyond V_T they are being retained on the column support by an affinity mechanism corresponding to $K>1$. The curve in Fig. 1 is commonly called a calibration curve. Methods used to generate the calibration curve will be discussed later. If a molecular weight calibration curve can be generated for the polymer of interest,

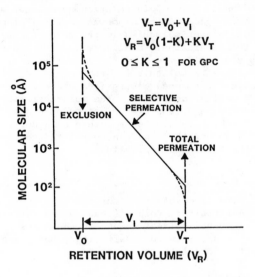

Figure 1.   Illustrative SEC Calibration Curve

then molecular weight statistics can be obtained by using the SEC raw trace in conjunction with this calibration curve. The fundamental aspects of the SEC separation mechanism have been treated theoretically by Casassa, et al,(2-6), Giddings (7), and Yau, et al,(8,9). These treatments are based on an equilibrium distribution of species between the mobile phase in the interstitial volume and the species in the pore volume of the column support.

Instrumentation

The essential components of the instrumentation are a solvent reservoir, a solvent delivery system (pump), sample injection system, packed columns, a detector(s), and a data processing system.

The heart of the instrumentaion is the fractionation column where the separation takes place. The most common packing material used has been a semi-rigid crosslinked polystyrene gel. Developments in column technology have made the low efficiency, large particle size (37-75µ) packing material obsolete. Currently, almost all the available SEC columns are packed with the high efficiency, microparticulate packings (<10µ). Recent state-of-the-art developments on column packings have been described by Majors (10). A listing of such type of packing materials is shown in Table 1.

The concentration of the polymer molecules eluting from SEC columns is continuously monitored by a detector. The most widely used detector in SEC is the differential refractometer (DRI), which measures the difference in refractive index between solvent and solute. Other detectors commonly used for SEC are functional group detectors; ultraviolet (UV) and infrared (IR), and absolute molecular weight detectors; low angle laser light scattering (LALLS) and in-line continuous viscometers. Applications of these detectors to SEC analysis will be discussed later in the Multiple Detectors Section. Other detectors also being used are the densimeter (11-19) and the mass detector (20-23).

Calibration

In order to convert a chromatogram into a molecular weight distribution curve, a calibration curve relating molecular weight to retention volume is required. Narrow MWD standards (polydispersity, M_w/M_n, is usually less than 1.1) of the polymer of interest are used to generate retention volume curves. A one to one correspondence of peak retention volume with peak molecular weight, M_p, is made. The peak retention volume is usually assigned to be $\sqrt{M_w \cdot M_n}$ for narrow MWD polymers. By plotting log M_p vs. retention volume, a primary molecular weight calibration curve is generated. The disadvantage of this method is that quite often well characterized narrow MWD polymer fractions of interest are not readily obtainable or require extensive laboratory time for their generation.

TABLE 1

Some Microparticulate Packings for SEC

Name	Supplier
(Semirigid Organic Gels)	
Finepak Gel	Jasco
Benson BN-X	Alltech, and Benson Co.
μ-Spheragel	Altex Scientific
Chromex	Altex Scientific
Shodex A	Perkin Elmer and Showa Denko
μ-Styragel	Waters Associates
PL Gel	Polymer Laboratories
BioBeads S	BioRad Laboratories
HSG	Shimadzu
MicroPak TSK type H, HXL	Toyo Soda and Varian
Ultrastyragel	Waters Associates
LiChrogel PS	EM Science
(Porous Silica Packings)	
LiChrospher	EM Science
Zorbax SE	DuPont
Zorbax PSM	DuPont
μ-Bondagel E	Waters Associates
μ-Porasil 60	Waters Associates
Glycophase-G	Pierce Chemical
MicroPak TSK Type SW	Toyo Soda & Varian

There are other methods for generating absolute MWD curves without resorting to polymer fractionation. One of these methods uses broad MWD standards to generate the molecular weight calibration curve (24-36). Other methods involve the use of the hydrodynamic volume concept. Polymers having different chemical structures or polymers having the same chemical structures but different chain configurations (linear vs. different types of branching) will have unique calibration curves. The SEC separation mechanism is based upon molecular size in solution (not molecular weight), or hydrodynamic volume. Therefore, if a parameter related to the hydrodynamic volume is used to generate calibration curves, a common calibration curve for a variety of polymers will be obtained. Benoit and coworkers (37) first proved the experimental validity of this concept by generating calibration curves consisting of a plot of the product of the intrinsic viscosity, $[\eta]$, and weight average molecular weight $\bar{M}_w$ vs. retention volume. With commercially available polystyrene standards such curves are readily generated. One can use

experimental and/or mathematical techniques (38) to obtain
secondary molecular weight calibration curves from the
hydrodynamic volume calibration curve as shown schematically in
Fig. 2.

Two recent refinements involving the use of hydrodynamic
calibration curves are: (1) Rudin's equation (39) which accounts
for the reduction of effective hydrodynamic volume of high
molecular weight polymers with finite concentration; (2) Hamielec
and Ouano's finding (40) that the hydrodynamic volume is the
product of intrinsic viscosity and M_n instead of M_w. This
refinement is of importance when applying hydrodynamic volume
considerations to molecular branching models for highly branched
and heterogeneous polymers. Transformation of the raw
chromatogram into various molecular weight averages, differential
and cumulative distribution curves was described by Pickett (41)
in one of his early papers. To numerically fit the calibration
curve, various approaches have been used, i.e., polynomial,
Yau-Malone equation (42) and a sum of exponentials. Detailed
discussion of these treatments can be found in Balke's book (43).

Instrument Spreading Correction

MWD curves calculated from SEC are generally broader than the
true or absolute MWD curves due to instrumental spreading of the
experimental chromatogram. Thus, the molecular weight averages
calculated from the experimental chromatograms can be
significantly different from the absolute molecular weight
averages. The instrument spreading in SEC has been attributed to
axial dispersion and skewing effects. Several computational
procedures (44-54) have been reported in the literature to
correct for these effects. In each method a specific shape for
the chromatogram of an ideal monodisperse species or narrow MWD
sample is assumed.

Tung (55) has shown that the normalized observed SEC
chromatogram, $F(v)$, at retention volume v is related to the
normalized SEC chromatogram corrected for instrument broadening,
$W(y)$, by means of the shape function $G(v,y)$ through the relation

$$F(v) = \int_{-\infty}^{\infty} G(v-y)w(y)dy \qquad (4)$$

Provder and Rosen (47) applying Tung's equation and the "Method
of Molecular Weight Averages" in conjunction with a linear
calibration curve derived the following two equations to obtain
the corrected values $M_n(c)$ and $M_w(c)$ from the uncorrected values
$\bar{M}_n(uc)$ and $\bar{M}_w(uc)$.

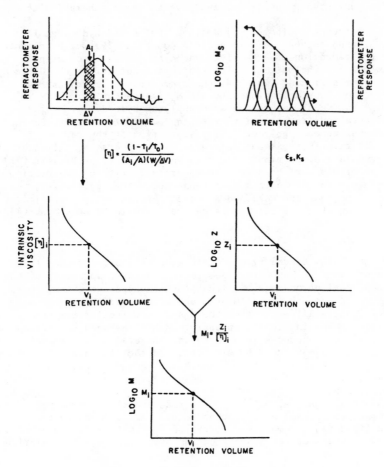

Figure 2.   Schematic Procedure for the Generation of Secondary
Molecular Weight Calibration Curve (Reprinted from Ref. 38.
Copyright 1973, American Chemical Society.)

$$\bar{M}_n(c) = \bar{M}_n(uc) \cdot \left[X_1 \cdot (1 + X_2) \right] \tag{5}$$

$$\bar{M}_w(c) = \bar{M}_w(uc) / \left[X_1 \cdot (1 - X_2) \right] \tag{6}$$

where

$$X_1 = 1/2 \left[[\bar{M}_n(t) / \bar{M}_n(uc)] + [\bar{M}_w(uc) / \bar{M}_w(t)] \right] \tag{7}$$

$$X_2 = (\phi - 1) / (\phi + 1) \tag{8}$$

$$\phi = [\bar{M}_n(t) \cdot \bar{M}_w(t)] / [\bar{M}_n(uc) \cdot \bar{M}_w(uc)] \tag{9}$$

$\bar{M}_n(t)$ and $\bar{M}_w(t)$ are the true or experimentally determined molecular weight averages. The "Method of Molecular Weight Averages" was included in the ASTM Standard Method, D3536-76, to correct for instrument spreading effects.

Multiple Detectors

Most size exclusion chromatographs use a DRI as a detector to monitor the concentration curves of samples eluting from the columns. This type of detector is highly sensitive and versatile and can monitor exceedingly low sample concentrations in a variety of solvents. However, it has several disadvantages which prevent it from being a "universal detector." At low and intermediate molecular weights, the specific refractive index increment at a given sample concentration is dependent upon the molecular weight. (56,57) For homopolymers, this difficulty can be circumvented by constructing a response factor curve vs. molecular weight. For multicomponent polymer systems, there is the additional complexity of the dependence of the specific refractive index increment upon the composition of the polymer system. In principle, if the structural features of the polymer system were known, response factor curves for a given multicomponent system could be constructed from a knowledge of atomic and bond refractions.(58) However, this is a very impractical approach for real polymer systems.
    Most coatings materials are complex multicomponent systems covering the low to intermediate molecular weight range. The use of a differential refractometer detector with the SEC provides useful routine screening information with regard to the approximate molecular weight distribution of these samples. However, little or no information can be inferred with regard to the compositional distribution as a function of molecular weight. To obtain this type of information on polymers, in the past SEC fractions have been collected and analyzed by infrared spectroscopy. In addition to being a tedious and time consuming method, a rather crude analysis of compositional distribution as a function of molecular weight is obtained.(59,60) To get maximum benefit from the SEC technique in terms of obtaining absolute molecular weight distributions and refined compositional distributions as a function of molecular weight, specific

functional group detectors coupled on-line to the SEC are required.

SEC/DRI/UV/IR   There have been a number of studies reported in the literature concerning the use of on-line functional group detectors (61-77) for SEC. The following examples show how SEC with multiple detectors can be used in qualitative analysis. Figure 3 shows an SEC/DRI/IR/UV chromatogram for a copolymer of methyl methacrylate (MMA) and vinyl acetate (VA) (25/75). Comparison of the SEC/IR trace with the SEC/DRI trace shows a difference in the ratio of the low retention volume to high retention volume peaks. The DRI detector has a different response to VA functionality than the MMA functionality at low retention volumes (high molecular weight). Although there are no UV active monomers present in the polymer, there is a UV detector response to the benzoyl peroxide initiator fragments attached to polymer chain ends. The difference in curve shape for the SEC/UV trace compared to the SEC/IR and SEC/DRI traces over the common retention volume range is indicative of a high degree of branching in this polymer. This is to be expected since vinyl acetate is known to produce branched polymers when made by emulsion polymerization techniques as was this copolymer of vinyl acetate. From this type of analysis of chain end distributions, valuable information about polymer chain-branching can be obtained.

Fig. 4 shows the SEC/UV/IR trace of a blend of a styrene/acrylic/acid terpolymer resin and a melamine resin. It is seen that there are three distinct peaks in the SEC/UV trace for this blend. The SEC/UV/IR traces show that the peak at ~185 ml corresponds to the polymer backbone; the middle peak at ~205 ml is associated with the melamine resin; and the third peak at ~220 ml has a strong UV absorbing characteristic and is acidic in nature and may well be caused by reaction by-products between catalyst, solvent and monomers. The melamine resin is melt blended with the terpolymer resin. This chromatogram indicates that only physical mixing occurs. The SEC/IR/UV information shown in this example is quite helpful in establishing proper blending conditions.

Fig. 5 shows the SEC/UV and SEC/IR traces of PMMA samples (78) which were photopolymerized with different concentrations of photosensitizer (0.05 x 10^{-2} M, 0.08 x 10^{-2} M, 0.25 x 10^{-2} M and 0.5 x 10^{-2} M). The photosensitizer used was 4,4' bis-(diethyl amino) benzophenone (DEABP). From the UV traces it is seen that the photosensitizers are chemically bound to the polymer chains. The results also seem to indicate that a greater number of sensitizer fragments reside in the lower molecular weight regions. A considerable amount of free sensitizer can be detected by the UV detector (retention volume ~210 ml) when the initial concentration of the sensitizer is above 0.08 x 10^{-2} M. The other auxiliary peaks beyond the retention volume of 200 ml could be due to some oligomeric components or solvent. The SEC/UV trace

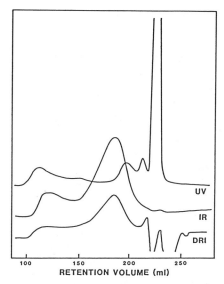

Figure 3.   SEC/DRI/IR/UV Chromatogram of Copolymer MMA/VA
(25/75)(Reprinted from ref. 69.   Copyright 1976 American Chemical
Society.)

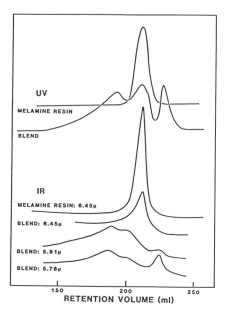

Figure 4.   SEC/UV/IR Chromatogram of a Blend of a
Styrene/Acrylic/Acid Terpolymer Resin and a Melamine Resin
(Reprinted from ref. 69.   Copyright 1976 American Chemical Society.)

at retention volumes less than 200 ml are polymer chains having
sensitizer fragments attached to the chain ends. Thus the UV
trace provides a distribution of polymer chain ends in these
samples. Values of M_n and M_w can be calculated by means of the
hydrodynamic volume approach. These results show that the
molecular weight of PMMA decreases with increasing concentration
of sensitizer. This is expected from the kinetics of conventional free-radical polymerization. Earlier results of the same
samples run on an SEC/THF instrument did not show this systematic
trend of molecular weights of PMMA as a function of DEABP
concentration. This is due to the fact that there was only a DRI
detector attached to the SEC/THF instrument. The DRI detector
picked up contributions from all the existing components which
may not be PMMA, such as those which show up at retention volumes
greater than 200 ml. These low molecular weight impurities
distorted the chromatograms with respect to molecular weight
distribution calculations. Consequently, the calculated
molecular weights and molecular weight distributions would be
erroneous. This illustrates one of the advantages of using the
SEC/IR traces. In addition, there are no negative peaks in the
SEC/IR traces as there are in the DRI trace. The absence of
these negative peaks allows much better definition of the low
molecular weight baseline cut-off point. Also, the IR-detector
is not as sensitive to room temperature fluctuations as is the
DRI and, therefore, the SEC/IR trace baseline will have better
long term stability. The same considerations with regard to
better baseline definition and long term stability apply to the
SEC/UV traces.

Quantitative Compositional-Molecular Weight Distribution
Considerations The previous examples demonstrated that crucial
qualitative information can be obtained about the composition of
components in multicomponent interpolymers and blends. Coatings
systems, typically, contain three to six components with some
present as minor constituents. To quantitatively determine the
compositional distribution as a function of molecular weight is a
rather formidable task for such complicated systems. In
addition, there are some complexities associated with using
multiple detectors for determining the compositional
heterogeneity of copolymer as discussed by Mori and Suzuki (73)
and Bressau (77). These complexities include accounting for: (a)
dead volume corrections, (b) hyperchromic shifts of copolymer
detection wavelengths, (c) variance of monomer component
absorptivity in the homopolymer to the copolymer, (d) validity of
the copolymer molecular weight scale or hydrodynamic volume
calibration approach and (e) mismatch of detector sensitivities
in either the low or high molecular weight ranges of the
chromatogram.
    In general the mathematical formulation of the quantitation
problem is shown in Table 2 for n-components. The matrix
equation, $H = AW \cdot G_p$, can be solved for the total polymer G_p and

weight fractions W_i in terms of the detector response factors A_{ij} by use of Cramer's rule. The rather complex mathematics does not lend itself to routine calculations of compositional weight fractions W_i or total polymer G_p unless a direct real-time on-line link to a powerful computer system is available. For example if 50 data points per sample are acquired for a three component system, it is necessary to solve a 3 x 3 matrix 50 times. For a copolymer system the equations become more tractable and lead to solutions for W_i and G_p as shown in Table 3. The two component case has been solved in the literature both by Adams (66) and by Runyon and coworkers (65) for a styrene-butadiene block copolymer. In these studies a DRI (detector 1) and UV (detector 2) detector were used in THF. The sytrene (component 1) and butadiene (component 2) contributed to the UV response resulting in $A_{22}=0$, thereby, simplifying equations for W_i and G_p in Table 3. Generally, in dealing with multi-component polymers, the detector system should be chosen so that simplifications can be made in the response factor matrix A, such that many off-diagonal A_{ij} elements will be equal to zero.

To experimentally determine the response factors A_{ij}, generally, the homopolymers of the components are monitored by the appropriate detector at several concentrations(62,65) The slope of the detector response (area under the appropriate SEC/detector trace) vs. concentration (grams), which should be linear, is then the response factor A_{ij}.

When the total polymer response, G_p is known as a function of retention volume, the molecular weight distribution can be obtained in the usual manner with the appropriate molecular weight calibration curve. The molecular weight calibration curve can be obtained: (a) by using the Runyon (65) copolymer molecular weight scale approach, or (b) by using a hydrodynamic volume approach if the Mark-Houwink constants for the polymer of interest are known or can be determined, or (c) by using a hydrodynamic volume approach in conjunction with an on-line viscosity detector.

SEC/LALLS   One of the absolute molecular weight detectors finding increasing usage is the low angle laser light scattering (LALLS) detector.(78-83) The unique features of the SEC/LALLS include: (a) simultaneous generation of the absolute molecular weight calibration curve and generation of the absolute molecular weight distribution by using a DRI and in conjunction with a LALLS, (b) being an excellent detector for aqueous SEC because it can generate an absolute molecular calibration curve, (c) its use for high temperature measurement, especially polyolefins, (d) being sensitive to very high molecular weight polymers, e.g. microgel. Fig. 6 is an example (82) of an NBS Standard SRM-1476 polyethylene in TCB at 135°C. It is seen that the LALLS detector clearly indicates a high molecular weight component that escapes detection with an IR detector. SEC/LALLS also has been used for

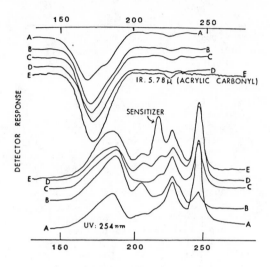

Figure 5.   SEC/UV/IR Chromatograms of Photopolymerized PMMA for [DEABP]: A is 5 x 10^{-4}M, B is 8 x 10^{-4}M, C is 12 x 10^{-4}M, D is 25 x 10^{-4}M, E is 50 x 10^{-4}M. (Reprinted from ref. 78. Copyright 1978 American Chemical Society.)

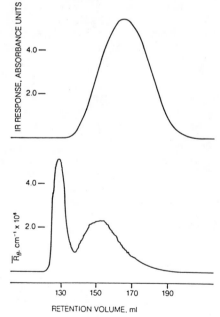

Figure 6.   SEC/LALLS Chromatogram of SRM 1476 in TCB at 135°C (Reprinted with permission from Ref. 82, Copyright 1979, John Wiley & Sons.)

TABLE 2

Quantitative Analysis of Multicomponent System

$$H_1 = (A_{11} W_1 + A_{12} W_2 + A_{13} W_3 + - - - A_{1N} W_N) \cdot G_P$$

$$H_2 = (A_{21} W_1 + A_{22} W_2 + A_{23} W_3 + - - - A_{2N} W_N) \cdot G_P$$

$$\vdots$$

$$H_N = (A_{N1} W_1 + A_{N2} W_2 + A_{N3} W_3 + - - - A_{NN} W_n \cdot G_P$$

$$\mathbf{H} = \mathbf{A W} \cdot G_P$$

$$\mathbf{H} = \begin{bmatrix} H_1 \\ H_2 \\ \vdots \\ H_N \end{bmatrix} ; \quad \mathbf{A} = \begin{bmatrix} A_{11} & A_{12} & - - - & A_{1N} \\ A_{21} & A_{21} & - - - & A_{2N} \\ \vdots & \vdots & & \vdots \\ A_{N1} & A_{N2} & - - - & A_{NN} \end{bmatrix} ; \quad \mathbf{W} \cdot G_P = \begin{bmatrix} W_1 \, G_P \\ W_2 \, G_P \\ \vdots \\ W_N \, G_P \end{bmatrix}$$

TABLE 3

Quantitative Analysis of Two-Component System

$$H_1 = \left[A_{11} W_1 + A_{12} W_2 \right] \cdot G_P$$

$$H_2 = \left[A_{21} W_1 + A_{22} W_2 \right] \cdot G_P$$

where,

H_1, H_2	– Detector Responses 1,2,
$A_{11}, A_{12}, A_{21}, A_{22}$	– Response Factors (Area/Gram),
G_P	– Weight in Grams of Polymer,
W_1, W_2	– Weight Fraction of Components 1,2,

$$W_1 = \frac{R \, A_{22} - A_{12}}{(A_{11} - A_{12}) + R \, (A_{22} - A_{21})} \quad ; \quad R = H_1 / H_2 ,$$

and

$$G_P = \left[\frac{H_1 (A_{21} - A_{22}) - H_2 (A_{11} - A_{12})}{A_{12} A_{21} - A_{22} A_{11}} \right]$$

the detection of shear degradation of polymers in SEC columns
(84), simultaneous calibration of molecular weight separation and
column dispersion (85), measurement of Mark-Houwink parameters
(86), determination of molecular weight and compositional
heterogeneity of block copolymers (87), and in obtaining
branching information in homopolymers and copolymers (88 - 98).
However, in using SEC/LALLS the analyst needs to be aware of some
data analysis considerations: (a) specific refractive index
increment, dn/dc, varies with molecular weight for low molecular
weight polymers and dependents on the composition of the
copolymers, (b) Virial coefficients depend on molecular weight.
(c) Transient noise spikes caused by bleeding of packing mateials
or passing of dust particles can occur, (d) There can be a
sensitivity mismatch between LALLS and DRI (e.g. inadequate
sensitivity in low molecular weight regions and detection of
microgel in high molecular weight regions), (e) Instrumental peak
broadening can occur in the scattering cell, (f) SEC/LALLS
provides only qualitative indications of polymer chain branching.

SEC/Viscometer Another on-line SEC detector which can provide
both absolute molecular weight statistics as well as branching
information is the viscosity detector. A discrete viscometry
technique (99-108) involving the coupling of a Ubbelohde-type
viscometer to measure the efflux time of each fraction was
reported in early 1970. The disadvantage of this type of
viscometer is that it is not a truly continuous detector. With
the speed and reduced column volumes and lower sample concentrations associated with modern high performance SEC this type of
viscometer detector is not practical. In 1972, Ouano (109)
developed a unique on-line viscometer which used a pressure
transducer to monitor the pressure drop across a capillary
continuously. More recently Lesec (110-112) and coworkers
described a similar and simpler on-line viscometer. In the
authors' laboratory, a differential transducer has been used to
monitor the pressure drop across a section of capillary tubing as
the polymer fractions elute from the SEC column. The experimental apparatus and performance evaluation were described
previously.(113,114) In 1984, the first commercially available
continuous viscosity detector for SEC was introduced by Viscotek
(115,116). The main component is the Wheatstone bridge
configuration consisting of four balanced capillary coils. Most
recently Abbott and Yau (117) described the design of a
differential pressure transducer capillary viscometer which is
comprised of two capillary tubes, one for eluting sample solution
and one for eluting solvent. The advantage of this device is
that the measured signal is independent of flow rate and
temperature fluctuations.

Like SEC/LALLS, the viscosity detector is sensitive to high
molecular weight fractions as shown in Fig. 7. A shoulder at
3,000,000 molecular weight detected by the DRI becomes a peak
when detected by the viscometer detector. The usefulness of the
SEC/Viscometer method is exemplified by the study of branched
polymers. Fig. 8 shows a log $[\eta]$ vs. log M_w plot for a randomly
branched polystyrene obtained from the SEC/Viscometer technique.

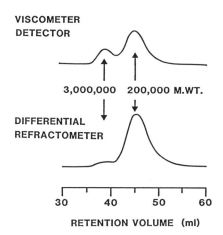

Figure 7.   DRI and Viscometer Chromatograms for a High
Molecular Weight PMMA Sample (Reprinted from ref. 114. Copyright
1984 American Chemical Society.)

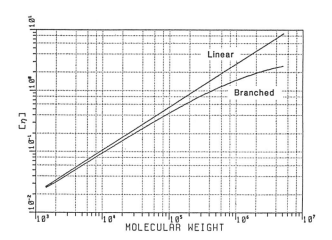

Figure 8.   Plot of log [η] <u>vs.</u> log $\bar{M}_w$ for a Randomly Branched
Polystyrene

The deviation from linearity in the high molecular weight region
can be clearly seen. Upon comparing the intrinsic viscosity with
that of the linear counterpart at the same molecular weight, the
branching index g' can be obtained as a function of molecular
weight. In addition, the SEC/Viscometer coupling can provide
absolute molecular weight averages, bulk intrinsic viscosity and
Mark-Houwink parameters from a single SEC experiment. A paper
dealing with the detailed description and the evaluation of the
data treatment for an SEC/Viscometer system can be found in this
volume. (118) An analyst using an SEC/Viscometer should be aware
of the following operational parameters which can produce errors
in the data: (a) flow variations caused by pump pulsations,
temperature fluctuations and restrictions in SEC columns as well
as in the connecting tubing, (b) mismatch between the flow rate
setting and actual delivered flowrate which can cause
concentration errors, (c) finite dead volume between detectors
which can produce a data offset between the DRI and viscometer
detector, (d) sensitivity mismatch between the viscometer and the
DRI detectors in the high and low molecular weight regions.

Oligomer Applications

The emergence of new coatings technologies such as high solids,
powder, water-borne and radiation curable coatings as a response
to governmental regulations has led to the development of resin
systems where the measurement of the oligomer and low molecular
polymer MWD is critically important in order to control the
properties of these coatings systems. Recently the HPSEC
technique, using high efficiency columns, has been shown to
provide the necessary resolution in the low molecular weight
region of interest for the above coatings systems. The high
efficiency columns result from the use of high pore volumes and
narrow particle size distribution of microparticulate packing
materials. The efficiency of a column is measured by plate
count. For a typical HPSEC column with 10 μm or less particle
packing, the plate count is usually in the order of 40,000
plates/m in contrast to about 1,500 plates/m for conventional
columns (37-75 μm particles). The ability of a column to
separate two adjacent peaks is expressed by the specific
resolution R_s as derived by Bly (119). For oligomer and small
molecule applications, R_s values are usually obtained from
various pairs of n-alkanes as reported in the literature
(120-122) for a variety of HPSEC columns from various vendors.
The effect of operational variables (e.g. flow rate, particle
size, column length, temperature, mobile phase, etc.) has been
studied by various groups (123-126). In general, the column
plate height decreases (efficiency increases) with decreasing
flow rate until an optimum flow rate is achieved in accordance
with the Van Deemter equation.(127) Consequently, to obtain high
resolution, the flow rate should be kept as low as possible. For
practical purpose, using THF as mobile phase, the flow rate is
usually set at 1 ml/min. The column efficiency also depends on
the particle size of the packings as shown by Vivilecchia and
coworkers.(124). Kato, et al (126) showed the effect of flow
rate, particle size and column length on the column plate count.

With the advent of high efficiency columns, HPSEC has become an indispensable chracterization and problem solving tool for oligomer analysis as will be shown in the following examples.(121,128,129)

HPSEC is very useful for screening various resins for the optimization of coatings viscosity and cured film properties. Among the five polyester resins shown in Fig. 9, E-17 was chosen to be scaled-up due to the unique combination of good film properties (hardness and salt spray resistance) and lowest viscosity. The three resins on the right hand side of Fig. 9 (E-44, E-38 & E-42) were not acceptable because their viscosities were too high as a result of high molecular weight components. While resin E-13 met the requirement of low viscosity for high solids, the film properties were not as good as those of E-17 due to the presence of a high level of unreacted monomer.

Figure 10 shows the HPSEC traces of two different batches of experimental acrylic resins. It is seen that due to the presence of high levels of low molecular weight components and residual monomer and solvent in sample TG-37, the Tg is 20°C lower than that of sample TG-57. Reducing the amount of low molecular weight components and residual monomer and solvent by vacuum stripping gave an increase in the Tg from 37°C to 48°C for sample TG-37. This brought the sample within the minimum acceptable Tg level consistent with non-"blocking" of the sample.

The most commonly used crosslinking agents for industrial coatings are melamine resins. Fig. 11 shows the HPSEC chromatograms of some melamine crosslinkers. M-1 is highly methylated and is claimed by the supplier to be monomeric, though at least four components are obviously present. M-2 is a partially methylated resin. It is claimed as polymeric, which is evidenced by the higher content of components in the higher molecular weight region. Due to the fact that it is only partially methylated M-2 has a higher tendency toward self-condensation. This phenomenon is demonstrated by comparing the MWD between the new and old M-2 resins as shown in the chromatograms. M-3 is a butylated resin and also is claimed to be monomeric by the supplier. This material also has a high tendency for self-condensation, presumably because of the steric hindrance of the bulky butyl group which interferes with further alkylation. The level of high molecular weight components in the melamine crosslinkers could be a direct reflection of the self-condensation reaction which would impart less impact resistance to a cured coating assuming all other parameters are the same.

In the production of epoxy esters, it becomes important to monitor changes in the molecular structure of low molecular weight epoxy resins during storage. It is known that catalyzed liquid epoxy resins will undergo further reaction upon aging. HPSEC has been used to monitor retains of incoming shipments from the resin supplier and monitor periodic samples from storage tanks of production plants. Figure 12 shows that at the time of sampling the samples that came from the plant storage tank were essentially similar to the retained samples from the supplier. Also shown in the figure is an epoxy sample which has been aged

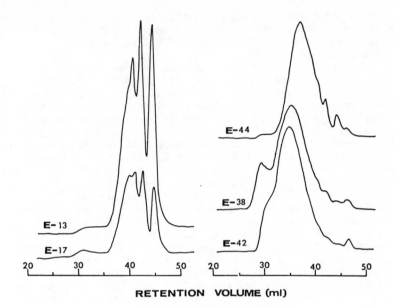

RETENTION VOLUME (ml)

Figure 9.   HPSEC Chromatograms of High Solids Polyesters
(Reprinted from ref. 128.   Copyright 1980 American Chemical
Society.)

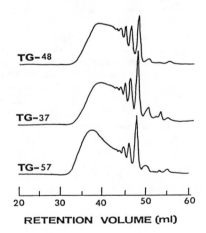

RETENTION VOLUME (ml)

Figure 10. HPSEC Chromatograms of Acrylic Resins (Reprinted
from ref. 128.   Copyright 1980 American Chemical Society.)

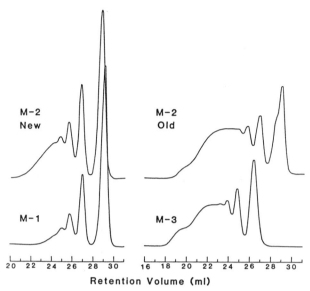

Figure 11.   HPSEC Chromatograms of Four Melamine Resins
(Reprinted with permission from Ref. 121, Copyright 1981, Marcel
Dekker, Inc.)

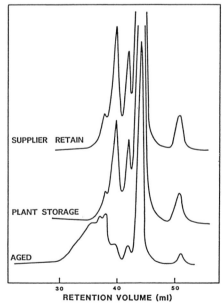

Figure 12.   HPSEC Chromatograms of Epoxy Resins (Reprinted
from ref. 128.   Copyright 1980 American Chemcial Society.)

for a year. It is seen that the low molecular weight components
had undergone further reaction to form a much higher molecular
weight compound. Observation of any changes in oligomer
distribution such as this, at any time, will alert the respective
production plant to take proper action.

Future Trends and Needs

In the area of column technology, the development and
commercialization of the columns for ultra high molecular weight
ranges (MW> 10^6) are needed. Also needed are very high
resolution columns for oligomers (50<MW<3000). With regard to
solvent delivery system, a pulseless pump having high accuracy
and high precision is critical, especially for an SEC/Viscometer
system. There is a need for enhanced sensitivity of oligomers
and small molecules. For SEC/UV, diode array spectrometry
providing a simultaneous multi-wavelength scan will be
advantageous for providing detailed compositional information for
polymers with UV active chromophores. Using a Fourier Transform
Infrared (FTIR) spectrometer for on-line polymer composition
determination and identification would expand SEC capability to a
significantly greater extent. Current literature available for
the application of FTIR to SEC in an on-line mode is quite
limited. (130-132) Besides the high cost of the FTIR, the main
obstacle is the availability of a suitable flow-through cell to
overcome mobile phase spectral interference and low solute
concentration. For complex polymers, the technique of the
orthogonal chromatography (133-135), should be explored. For
oligomers, supercritical fluid chromatography (SFC) (136-140) has
a great potential. Chromatographic methods for gel content
determination now are more feasible with LALLS and viscometric
detectors and should be re-examined. (141-145).
     This overview covers only non-aqueous SEC. For the theory,
practice, and applications of aqueous SEC, the reader is referred
to the literature (146-155) as well as Barth's chapter (156) in
this volume. For general references on SEC, the reader is
referred to the published monographs (157-168).

Summary

This paper has presented a review of the SEC separation
mechanism, molecular weight calibration methods and instrument
spreading correction methods. Examples were shown for the
application of multiple detectors to the determination of
absolute molecular weight distribution of polymers, compositional
distribution as a function of molecular weight of copolymers and
branching information for non-linear polymers. Examples also
were shown how HPSEC can be used for guiding polymer synthesis
and processing, correlating oligomer distribution with end-use
properties and monitoring the quality of supplier raw materials.
In recent years, HPSEC has become an indispensable
characterization and problem-solving tool for the analysis of
oligomers and polymers in the plastics, rubbers,and coatings
industries. The information generated by means of the HPSEC

technique has significantly aided polymer chemists and coatings
formulators to tailor-make coatings systems to meet specific
end-use properties.

Literature Cited

1. Moore, J. C., J. Polym. Sci., A-2, 2, 285(1964).
2. Casassa, E. F., J. Polym. Sci., Part B, 5, 773(1967).
3. Casassa, E. F. and Tamami, Y., Macromolecules, 2, 14(1969).
4. Casassa, E. F., Separ. Sci., 6, 305(1971).
5. Casassa, E. F., J. Phys. Chem., 75, 3929(1971).
6. Casassa, E. F., Macromolecules, 9, 182(1976).
7. Giddings, J. C., Kucera, E., Russell, C. P. and Myers, M.
   N., J. Phys. Chem., 78, 4397(1968).
8. Yau, W. W., Malone, C. P. and Fleming, S. W., J. Polym.
   Sci., Part B, 6, 803(1968).
9. Yau, W. W. and Malone, C. P., Polym. Prep., Am. Chem. Soc.,
   Div. Polym. Chem., 12(2) 797(1971).
10. Majors, R. E., J. Chromatogr. Sci., 18, 488(1980).
11. Leopold, H. and Trathnigg, B., Angew. Makromol. Chem., 68,
    185(1978).
12. Francois, J., Jacob, M., Grubisic-Gallot, Z. and Benoit, H.,
    J. Appl. Polym. Sci., 22, 1159(1978).
13. Trathnigg, B. and Leopold, H., Makromol. Chem., Rapid
    Commun., 1, 569(1980).
14. Trathnigg, B., Makromol. Chem., 89, 73(1980).
15. Kratky, O., Leopold, H. and Stabinzer, H., Angew Phys. 7,
    273(1969).
16. Trathnigg, B., Angew Makromol. Chem., 89, 65(1980).
17. Trathnigg, B. and Jorde, C., J. Chromatogr., 24, 147(1982).
18. Elsdon, W. L., Goldwasser, J. M. and Rudin, A., J. Polym.
    Sci., Polym. Chem. Ed., 20, 3271(1982).
19. Boyd, D., Narasimlan, V., Huang, R. Y. M. and Burns, C. M.,
    J. Appl. Polym. Sci., 29, 595(1984).
20. Charlesworth, J. M., Anal. Chem., 50, 1414(1978).
21. Applications Report, Applied Chromatography Systems, Ltd.
22. Morris, E. E. M. and Grabovac, I., J. Chromatogr., 189,
    259(1980).
23. Grinshpun, V. and Rudin, A., J. Appl. Polym. Sci., 32,4303
    (1986).
24. Cantow, J. J. R., Porter, R. S. and Johnson, J. F., J.
    Polym. Sci. A-1, 5, 1391(1967).
25. Weiss, A. R. and Cohn-Ginsberg, E., J. Polym. Sci. A-2, 8,
    148(1970).
26. Wild, L., Ranganath, R. and Ryle, T. J. Polym. Sci., A-2,
    9, 2137(1971).
27. Swartz, T. D., Bly, D. D. and, Edwards, A. S., J. Appl.
    Polym. Sci., 16, 3353(1972).
28. Abdel-Alim, A. H. and Hamielec, A. E., J. Appl. Polym. Sci.,
    18, 297(1974).
29. Balke, S. T., Hamielec, A. E., LeClair, B. P. and Pearce, S.
    L., Ind. Eng. Chem. Prod. Res. Dev., 8, 54(1969).
30. Frank, F. C., Ward, I. M. and Williams, T., J. Polym. Sci.
    A-2, 6, 1357(1968).

31. Friis, N. and Hamielec, A. E., Advances in Chromatography
    (Giddings, J. C., Grushka, E., Keller, R. A. and Cazes, J.
    eds.), 13, 41(1975).
32. Yau, W. W., Stoklosa, H. J. and Bly, D. D., J. Appl. Polym.
    Sci., 21, 1911(1977).
33. Provder, T., Woodbrey, J. C. and Clark, J. H., Separ. Sci.,
    6, 101(1971).
34. Hamielec, A. E. and Omorodion, S. N. E., ACS SYMPOSIUM
    SERIES No. 138 (Provder, T. ed.) 183(1980).
35. Kubin, M., J. Appl. Polym. Sci., 27, 2933(1982).
36. Kubin, M., J. Liq. Chrmatogr., 7(S-1), 41(1984).
37. Benoit, H., Grubisic, Z. and Rempp, P., J. Polym. Sci., Part
    B, 5, 753 (1967).
38. Provder, T., Clark, J. H. and Drott, E. E., Advances in
    Chemistry Series, 125, 117(1973).
39. Rudin, A. and Wagner, R. A., J. Appl. Polym. Sci., 20,
    1483(1976).
40. Hamielec, A. E. and Ouano, A. C., J. Liq. Chromatog., 1,
    111(1978).
41. Pickett, H. E., Cantow, M. J. R. and Johnson, J. F., J.
    Appl. Polym. Sci., 10, 917(1966).
42. Yau, W. W. and Malone, C. P., J. Polym. Sci., Polym. Letters
    Ed., 5, 663(1967).
43. Balke, S. T. in "Quantitative Column Liquid Chromatography;
    a Survey of Chemometric Methods", Elsevier, NY, 1984,
    pp.204.
44. Hess, M. and Kratz, R. F., J. Polym. Sci., A-2, 4, 731
    (1966).
45. Pickett, H. E., Cantow, M. J. R. and Johnson, J. F., J.
    Polym. Sci., C-21, 67(1968).
46. Balke, S. T. and Hamielec, A. E., J. Appl. Polym. Sci., 13,
    1381(1969).
47. Provder, T. and Rosen, E. M., Separ. Sci., 5, 437(1970).
48. Halvorson, H. R. and Ackers, G. K., J. Polym. Sci., A-2, 9,
    245 (1971).
49. Pierce, P. E. and Armonas, J. E., J. Polym. Sci., C-21, 23,
    (1968).
50. Timm, D. C. and Rachow, J. W., J. Polym. Sci., Polym. Chem.
    Ed., 13, 1401 (1975).
51. Biesenberger, J. A. and Ouano, A., J. Appl. Polym. Sci., 14,
    471 (1970).
52. Adesanya, B. A., Yen, H. C., Timm, D. C. and Plass, N. C.,
    Organic Coatings and Applied Polymer Science Proceedings,
    48, 870 (1983).
55. Tung, L. H., J. Appl. Polym. Sci., 10, 375 (1966).
56. Barrall, E. M., Cantow, M. J. R. and Johnson, J. F., J.
    Appl. Polym. Sci., 12, 1373 (1968).
57. Law, R. D., J. Polym. Sci., A-1, 7, 2097 (1969).
58. Daniels, F. and Alberty, R. A., "Physical Chemistry", John
    Wiley & Sons, Inc., N. Y., 1966, p. 674.
59. Bartosiewics, R. L., J. Paint Tech., 39, 28 (1967).
60. Anderson, D. G. and Isakson, K. E., Polymer Preprints,
    11(2), 1190 (1970).

61. Rodriguez, F., Kulakowski, R. A. and Clark, O. K., I & EC Product Research and Development, 5, 121 (1966).
62. Terry, S. L. and Rodriguez, F., J. Polym. Sci., C-21, 191 (1968).
63. Ross, J. H. and Casto, M. E., ibid, C-21, 143 (1968).
64. Ross, J. H. and Shank, R. L., Polymer Preprints, 12, (3), 812 (1971).
65. Runyon, J. R., Barnes, D. E., Rudd, J. F. and Tung, L. H., J. Appl. Polym. Sci., 13, 2359 (1969).
66. Adams, H. E., Separ. Sci., 6(2), 259 (1971).
67. Chang, F. S., J. Chromatogr., 55 67 (1971).
68. Mirabella, Jr., F. M., Barrall, II, E. M. and Johnson, J. F., J. Appl. Polym. Sci., 20, 581 (1976).
69. Provder, T. and Kuo, C., Organic Coatings and Plastics Chemistry Preprints, 36(2), 7 (1976).
70. Stojanov, Ch., Shirazi, Z. H. and Audu, T. O. K., Chromatographia, 11, 63(1978).
71. Stojanov, Ch., Shirazi, Z. H. and Audu, T. O. K., Chromatographia, 11, 274(1978).
72. Mori, S., J. Chromatogr., 157, 75(1978).
73. Mori, S. and Suzuki, T., J. Liq. Chromatogr., 4, 1685(1981).
74. Elgert, K. F. and Wohlschiess, R., Agnew Makromol. Chem., 51, 87(1977).
75. Balke, S. T. and Patel, R. D., in "Size Exclusion Chromatography (GPC)"; Provder, T., Ed., ACS SYMPOSIUM SERIES, No. 138, 1980; pp. 149-82.
76. Garcia-Rubio, L. H., Hamielec, A. E. and MacGregor, J. F., ACS SYMPOSIUM SERIES, No. 203, C. D. Craver, Ed., 1983, pp. 310.
77. Bressau, R., in "Liquid Chromatography of Polymers and Related Materials II", Cazes, J.; Delamare, X., Eds., Marcel Dekker, NY, 1979, pp.73.
78. McGinniss, V. D., Provder, T., Kuo, C. and Gallopo, A., Macromolecules, 11, 393 (1978).
79. Ouano, A. C. and Kaye, W., J. Polym. Sci., Polym. Chem. Ed., 12 1151 (1974).
80. McConnell, M. L., Am. Lab., 10(5), 63 (1978).
81. Jordan, R. C., J. Liquid Chromatogr., 3, 439 (1980).
82. MacRury, T. B. and McConnell, M. L., J. Appl. Polym. Sci., 24, 651 (1979).
83. Malihi, F., Kuo, C. and Provder, T., J. Appl. Polym. Sci., 29, 925 (1984).
84. Rooney, J. G. and Ver Strate, G., in "Liquid Chromatography of Polymers and Related Materials III", Cazes, J., Ed., Marcel Dekker, N.Y. 1981, pp. 207.
85. He, Z.-D., Zhang, X.-C. and Cheng, R.-S., J. Liq. Chromatography, 5, 1209 (1982)
86. Grinshpun, V. and Rudin, A., Makromel. Chem. Rapid Commun., 6, 219 (1985).
87. Dumelow, T., Holding, S.R., Maisey, L. J. and Dawkins, J. V., Polymer, 27, 1170 (1986).
88. Hamielec, A. E., Ouano, A. C. and Nebenzahl, L. L., J. Liq. Chromatogr., 1, 527 (1978).

89. Jordan, R. C. and McConnell, M. L., in "Size Exclusion
    Chromatography (GPC)", Provder, T., Ed., ACS SYMPOSIUM
    SERIES, Vol. 138, 1980, pp.107-129.
90. Axelson, D. E. and Knapp, W. C., J. Appl. Polym. Sci., 25,
    119 (1980).
91. Agarwal, S. H., Jenkins, R. J. and Porter, R. S., J. Appl.
    Polym. Sci., 27, 113 (1982).
92. Jordan, R. C., Siler, S. F., Sehon, R. D. and Rivard, R. J.,
    in "Size Exclusion Chromatography Methodology and
    Characterization of Polymers and Related Materials",
    Provder, T., Ed., ACS SYMPOSIUM SERIES, No. 245, 295 (1984).
93. Roovers, J. and Toporowski, P. M., Macromolecules, 14,
    1174(1981).
94. Roovers, J., Hadjichristidis, N. and Fetters, L. J.,
    Macromolecules, 16, 214(1983).
95. Rudin, A., Grinshpun, V. and O'Driscoll, K. F., J. Liq.
    Chromatogr., 7, 1009(1984).
96. Grinshpun, V, Rudin, A., Russell, K. E. and Scammell, M. V.,
    J. Polym. Sci., Polym. Phys. Ed., 24, 1171 (1986).
97. Hamielec, A. E., Meyers, H., in "Developments in Polymer
    Characterization - 5", Dawkins, J. V., Ed., Applied Science
    Publishers, NY (1986) pp.95.
98. Hjertberg, T., Kulin, L.-I. and Sorrik, E., Polym. Testing,
    3, 267(1983).
99. Meyerhoff, G., Makromol. Chem., 118, 265(1968).
100. Goedhart, D., Opschoor, A., J. Polym. Sci., A-2, 8, 1227
     (1970).
101. Meyerhoff, G., Separ. Sci., 6, 239(1971).
102. Grubisic-Gallot, Z., Picot, M., Gramain, P. and Benoit, H.,
     J. Appl. Polym. Sci., 16, 2931 (1972).
103. Gallot, Z., Marais, L. and Benoit, H., J. Chromatogr., 83,
     363 (1973).
104. Servotte, A. and DeBruille, R., Makromol. Chem., 176,
     203(1975).
105. Park, W. S. and Graessley, W. W., J. Polym. Sci., Polym.
     Phys. Ed., 15, 71(1977).
106. Constantin, D., Eur. Polym. J., 13, 907(1977).
107. Janca, J. and Kolinsky, S. J., J. Chromatogr., 132,
     187(1977).
108. Janca, J. and Pokorny, S., J. Chromatogr., 134, 273(1977).
109. Ouano, A. C., J. Polym. Sci., A-1, 10, 2169 (1972).
110. Lesec, J. and Quivoron, C., Analusis, 4, 399(1976).
111. Letot, L., Lesec, J. and Quivoron, C., J. Liq. Chromatogr.,
     3, 427(1980).
112. Lecacheux, D., Lesec, J. and Quivoron, C., J. Appl. Polym.
     Sci., 27, 4867(1982).
113. Malihi, F. B., Koehler, M. E., Kuo, C. and Provder, T.,
     Pittsburgh Conference on Analytical Chemistry and Applied
     Spectroscopy, Paper No. 806(1982).
114. Malihi, F. B., Kuo, C., Koehler, M. E., Provder, T. and Kah,
     A. F., "Size Exclusion Chromatography Methodology and
     Characterization of Polymers and Related Materials",
     Provder, T., Ed., ACS SYMPOSIUM SERIES, No. 245, 281(1984).
115. Viscotek Corporation, Porter, Texas.

116. Haney, M. A., J. Appl. Polym. Sci., 30, 3037(1985).
117. Abbott, S. D. and Yau, W. W., U. S. Patent 4, 578, 990, April 1, 1986.
118. Kuo, C., Provder, T., Koehler, M. E. and Kah, A. F., "Use of a Viscometer Detector for SEC Characterization of MWD and Branching in Polymers", this volume.
119. Bly, D. D., J. Polym. Sci., Part C., 21, 13 (1968).
120. Krishen, A., and Tucker, R. G., Anal. Chem. 99, 898 (1977).
121. Kuo, C., Provder, T., Holsworth, R. M. and Kah, A. F., Chromatogr. Sci. Series, 19, 169 (1981).
122. Chiantore, O. and Guaita, M., J. Liq. Chromatogr., 9, 1341 (1986).
123. Little, J. N., Waters, J. L., Bombaugh, K. J. and Pauplis, W. J., Separ. Sci., 5, 765 (1970).
124. Vivilecchia, R. V., Cotter, R. L., Limpert, R. J., Thimot and N. Z. Little, J. N., J. Chromatogr., 99, 407 (1974).
125. Yau, W. W., Kirkland, J. J., Bly, D. D. and Stoklosa, H. J., J. Chromatogr., 125, 219 (1976).
126. Kato, Y., Kido, S., Yamamoto, M. and Hasimoto, T., J. Polym. Sci., Polym. Phys. Ed., 12, 1339 (1974).
127. Van Deemter, J. J., Zuiderweg, F. J. and Klinkenberg, A., Chem. Eng. Sci., 5, 271 (1956).
128. Kuo, C. and Provder, T., in "Size Exclusion Chromatography (GPC)", Provder, T., Ed., ACS SYMPOSIUM SERIES, Vol. 138, 1980, pp. 207-224.
129. Kuo, C., Provder, T. and Kah, A. F., Paint & Resin, 53(2), 26(1983).
130. Vidrine, D. W., J. Chromatogr. Sci., 17, 477(1979).
131. Brown, R. S., Hausler, D. W., Taylor, L. T. and Carter, R. C., Anal. Chem., 53, 197(1981).
132. Folster, U. and Herres, W., Farbe & Lack, 89, 417(1983).
133. Balke, S. T., Sep. Purif. Methods, 11, 1(1982).
134. Balke, S. T. and Patel, R. D., Adv. Chem. Series, 203, 281(1983).
135. Balke, S. T., This Volume.
136. Jentoft, R. E. and Fouw, T. H., J. Polym. Sci., Polym. Letters, 7, 811(1969).
137. Schmitz, F. P. and Klesper, E., Polym. Bull., 5, 603(1981).
138. Fjeldsted, J. C., Jackson, W. P., Peaden, P. A. and Lee, M. L., J. Chromatogr. Sci., 21, 222(1983).
139. Hirata, Y. and Nakata, F., J. Chromatogr., 295, 315(1984).
140. White, C. M. and Houck, R. K., HRC&CC, 9, 4(1986).
141. Gaylor, V. F., James, H. L. and Herdering, J. P., J. Polym. Sci., 13, 1575 (1975).
142. Williamson, T. J., Gaylor, V. F. and Piirma, I., ACS Symposium Series, 138, 77 (1980).
143. Hellman, M. Y., Bowmer, T. and Taylor, G. N., ACS Organic Coatings and Plastics Chemistry Preprints, 45, 126 (1981).
144. Malihi, F. B., Kuo, C. and Provder, T., J. Liq. Chromatogr. 6, 667 (1983).
145. Trowbridge, D., Brower, L., Seeger, R. and McIntyre, D., This Volume.
146. Barth, H. and Regnier, F., J. Chromatogr., 192, 275(1980).
147. Barth, H., Anal. Biochem., 124, 191(1982).

148. Kim, C. J., Hamielec, A. E. and Benedek, A., J. Liq.
     Chromatogr., 5, 425(1982).
149. Cooper, A. R. and Van DerVeer, D. S., J. Liq. Chromatogr.,
     1, 693 (1978).
150. Rollings, J. E., Bose, A., Caruthers, J. M., Tsao, G. T. and
     Okos, M. R., in "Polymer Characterization Spectroscopic,
     Chromatographic, and Physical Instrumental Methods", Craver,
     C. D., Ed., ADV. CHEM. SERIES, No. 203, 1983, pp.345.
151. Levy, I. J. and Dubin, P. L., Ind. Eng. Chem. Prod. Res.
     Dev., 21, 59(1982).
152. Dubin, P. L., Am. Lab., Jan., 60(1983).
153. Muller, G. and Yonnet, C., Makromol. Chem., Rapid Commun.,
     5, 197(1984).
154. Omorodin, S. N. E., Hamielec, A. E. and Brash, J. L., ACS
     SYMPOSIUM SERIES, No. 138, 267(1980).
155. Hashimoto, T., Sasaki, H., Aiura, M. and Kato, Y., J. Polym.
     Sci., Polym. Phys. Ed., 16, 1789(1978).
156. Barth, H., in this volume.
157. Balke, S. T., "Quantitative Column Liquid Chromatography - A
     Survey of Chemometric Methods," Elsevier, N.Y. (1984).
158. Provder, T., Ed., "Size Exclusion Chromatograpy -
     Methodology and Characterization of Polymers and Related
     Materials", ACS SYMPOSIUM SERIES, No. 245, American Chemical
     Society, Washington, D.C. (1984).
159. Janca, J., Ed., "Steric Exclusion Liquid Chromatography of
     Polymers", CHROMATOGRAPHIC SCIENCE SERIES, Vo. 25, Marcel
     Dekker, Inc., N.Y. (1984).
160. Cazes, J., Ed., "Liquid Chromatography of Polymers and
     Related Materials III", CHROMATOGRAPHIC SCIENCE SERIES, Vol.
     19, Marcel Dekker, Inc., N.Y. (1981).
161. Cazes, J. and Delamare, V., Eds., "Liquid Chromatography of
     Polymers and Related Materials II", CHROMATOGRAPHIC SCIENCE
     SERIES, Vol. 13, Marcel Dekker, Inc., N.Y. (1980).
162. Provder, T., Ed., "Size Exclusion Chromatography (GPC)", ACS
     SYMPOSIUM SERIES, No. 138, American Chemical Society,
     Washington, D.C. (1980).
163. Yau, W. W., Kirkland, J. J. and Bly, D. D., "Modern Size
     Exlusion Liquid Chromatography", Wiley-Interscience, N.Y.
     (1979).
164. Cazes, J., Ed., "Liquid Chromatography of Polymers and
     Related Materials I", CHROMATOGRAPHIC SCIENCE SERIES, Vol.
     8, Marcel Dekker, Inc., N.Y. (1977).
165. Ezrin, M., Ed., "Polymer Molecular Weights", ADVANCES IN
     CHEMISTRY SERIES, No. 125, American Chemical Society,
     Washington, D. C. (1973).
166. Altgelt, K. H. and Segal, L., Eds., "Gel Permeation
     Chromatography", Marcel Dekker, Inc., N.Y. (1971).
167. Determann, H., "Gel Chromatography", Springer Verlag, Berlin
     (1968).
168. Johnson, J. F. and Porter, R. S., Eds., "Analytical Gel
     Permeation Chromatography", Journal of Polymer Science Part
     C, Polymer Symposia No. 21, N.Y. (1967).

RECEIVED June 26, 1987

Chapter 2

Nonsize Exclusion Effects in High-Performance Size Exclusion Chromatography

Howard G. Barth

Hercules Inc., Research Center, Wilmington, DE 19894

This chapter presents an overview of nonideal size
exclusion chromatographic (SEC) behavior which may
occur during high performance SEC. If not eliminated
or at least reduced, these effects may lead to
erroneous molecular weight distribution results.
Enthalpic interactions between polymer and packing
are discussed in detail as well as intramolecular
electrostatic effects that occur with polyelectrolytes. Concentration effects, that is,
viscous fingering and macromolecular crowding, are
reviewed. Other nonsize exclusion effects, which may
exist especially in high performance systems, are
presented including polymer shear degradation,
ultrafiltration, and polymer chain orientation.

Size exclusion chromatography is a unique separation technique
based on molecular size (hydrodynamic volume) differences among
solutes. The distribution coefficient K_d of an eluting solute is
defined as

$$K_d = (V_e - V_o)/V_i \qquad (1)$$

where K_d is the ratio of the solute concentration within the
pores of the packing to the solute concentration in the
interstitial volume, V_e is the elution volume of the polymer, and
V_o and V_i are the interstitial and pore volumes of the packed
column, respectively.
    The physical significance of the values that K_d can have
is as follows:

$K_d = 0$ Complete exclusion of the solute from the
pores of the packing.

$0<K_d<1$ Partial exclusion; the solute distributes
between V_i and V_o, but the average
solute concentration is higher in V_o.

0097–6156/87/0352–0029$06.00/0
© 1987 American Chemical Society

$K_d = 1$ Total permeation; the solute freely permeates into and out of the pores of the packing; the average solute concentration in V_i and V_o are equal.

$K_d > 1$ Enthalpic interactions between solute and packing are occurring.

Theory. The most widely accepted mechanism of size separation is based on steric exclusion (1). In terms of thermodynamic properties, the distribution coefficient consists of enthalpic and entropic contributions:

$$K = e^{\Delta S^\circ/R}\ e^{-\Delta H^\circ/RT} \qquad (2)$$

where ΔS° is the loss of conformational entropy when 1 mole of solute passes from V_o to V_i and ΔH° is the energy released when 1 mole of solute interacts with the packing (2).

Under ideal SEC conditions, whereby $\Delta H = 0$ and equilibrium is attained:

$$K_d = e^{\Delta S^\circ/R} \qquad (3)$$

Thus the distribution coefficient that one obtains from an SEC experiment (eqn. 1) should only be a function of the change in conformational entropy of the solute, which is dependent on the size and shape of the solute with respect to the size and shape of the pores of the packing. The SEC distribution coefficient is also independent of temperature, unlike all other chromatographic techniques. In practice, however, K_d is marginally dependent on temperature because of conformational changes that the macromolecule may experience with changing temperature. Also, the pore structure and size of the packing, especially for soft gels, may be affected by temperature, which in turn will influence K_d. Column efficiency will increase with increasing temperature; this will indirectly affect the K_d of the eluting species through a decrease in band broadening.

To take into account nonideal SEC behavior, Dawkins (3-7) introduced the distribution coefficient K_p:

$$K_p = e^{-\Delta H^\circ/RT} \qquad (4)$$

If enthalpic interactions are present, the distribution coefficient K_d obtained from an SEC experiment can be defined as

$$K_d = K_p K_d \qquad (5)$$

or substituting into eqn. 1 and rearranging:

$$V_e = V_o + K_p K_d V_i \qquad (6)$$

Therefore, for a separation based solely on steric exclusion, K_p is unity.

Figure 1 shows the relationship between SEC and all other liquid chromatography modes (8). In SEC we are dealing with a relatively small working volume V_i, hence peak capacity is highly limited as compared to interactive chromatographic separations (8). It is also of interest to note that an SEC column can function as an interactive column, and vice versa; this can be done by promoting enthalpic interactions through changes in mobile phase composition. For example, Kopaciewicz and Regnier (9) explored the use of silica-based packings to separate proteins under nonideal SEC conditions. They found that proteins could be selectively adsorbed, ion excluded, or size excluded by varying mobile phase pH relative to the isoelectric point of the protein. Mori and Yamakawa (10) used polystyrene packing as both an SEC and normal-phase adsorption column by utilizing chloroform and chloroform-hexane mobile phases, respectively.

Nonsize-Exclusion Effects. To develop a reliable SEC method, one must not only ensure that enthalpic interactions are zero, but also must take into account or eliminate the following nonsize-exclusion effects that will lead to nonideal SEC behavior:

• Solute/packing enthalpic interactions.

• Intramolecular electrostatic effects.

• Concentration effects.

• Polymer shear degradation.

• Ultrafiltration.

• Hydrodynamic effects.

• Polymer chain orientation and deformation.

• Peak dispersion.

Another type of nonideal SEC behavior, which will not be covered in this chapter, is related to the use of mixed mobile phases (multiple solvents). Because solute-solvent interactions play a critical role in controlling the hydrodynamic volume of a macromolecule, the use of mixed mobile phases may lead to deviations from ideal behavior. Depending on the solubility parameter differences of the solvents and the solubility parameter of the packing, the mobile phase composition within the pores of the packing may be different from that in the interstitial volume. As a result, the hydrodynamic volume of the polymer may change when it enters the packing leading to unexpected elution results. Preferential solvation of the polymer in mixed solvent systems may also lead to deviations from ideal behavior (11).

With the introduction of high performance columns, in which

high shear forces are present, polymer shear degradation and chain orientation are more of a concern as compared to conventional SEC packings. Also because of the small interstices present in a high performance column, ultrafiltration may lead to incomplete polymer elution. However, with the high plate counts and thus excellent resolution of high performance columns, peak dispersion corrections are not as important as they are for conventional packings of much lower efficiencies. Although, as we shall see, concentration effects can be more pronounced with high efficiency columns.

Nonsize–exclusion effects caused by the use of small diameter packings are highly dependent on the molecular weight of the polymer and, in fact, severely limit the use of SEC for the analysis of ultrahigh molecular weight polymers as discussed in detail by Giddings (12).

In this chapter, we will present an overview of these nonideal-SEC effects, with the exception of peak dispersion which has been extensively covered by others (for example, see references 13 and 14). The reader should also consult references 8 and 13-19 for general background material and reviews on this subject.

Enthalpic Interactions

There are a number of different enthalpic interactions that can occur between polymer and packing, and in many cases multiple interactions can exist depending on the chemical structure of the polymer. Enthalpic interactions that are related to water–soluble polymers include ion exchange, ion inclusion, ion exclusion, hydrophobic interactions, and hydrogen bonding (17). Other types of interactions commonly encountered in SEC, as well as in all other chromatographic separations, are dispersion (London) forces, dipole interactions (Keeson and Debye forces), and electron-donor-acceptor interactions (20).

To eliminate these effects, a mobile phase is chosen that is a good solvent for the polymer and whose solubility parameter, δ, is close to that of the packing. Thus the polymer and packing are well solvated and potential adsorptive sites on both are "deactivated". As demonstrated by Dawkins (21,22) and Mori (10,23), if $\delta_{gel} > \delta_{solvent}$, normal–phase adsorption will occur. If $\delta_{gel} < \delta_{solvent}$, the packing will act as a reversed–phase packing. At $\delta_{gel} = \delta_{solvent}$, size exclusion is the dominant separation mechanism. For crosslinked polystyrene, which is the most commonly used SEC packing, it is best to use mobile phases that have solubility parameters of approximately 9.1, such as chloroform, tetrahydrofuran, or toluene. In the case of other packings, such as silica, solvent selection schemes commonly used in HPLC can be used to eliminate adsorption. Silica packings can also be surface modified to help eliminate adsorption.

In addition to the solubility parameter model to treat SEC adsorption effects, an approach based on Flory–Huggins interaction parameters has also been proposed (24-27). For an excellent review of both mechanisms, see reference 28. A general treatment of polymer adsorption onto chromatographic packings can be found in Belenkii and Vilenchik's recent book (29).

Adsorption. Hydrophobic interactions, which may occur using aqueous mobile phases, usually can be eliminated by the addition of an organic modifier to the aqueous mobile phase (30,33) or by a reduction of ionic strength (34,35). Recently, Haglund and Marsden (36–40) have undertaken a systematic study on the chromatographic behavior of low molecular weight solutes on Sephadex packings and explained these results in terms of hydrophobic interactions.

Hydrogen bonding, which can be prevalent in both aqueous and nonaqueous SEC systems especially with the use of silica packings, can be eliminated by the addition of urea or guanidine to the mobile phase (41) or the use of aprotic solvents, such as dimethylsulfoxide, dimethylformamide, or formamide, which act as hydrogen bond acceptors. The use of aprotic solvents not only helps in eliminating solute-packing interactions, but also prevents polymer association caused by intramolecular hydrogen bonding as in the case of polysaccharides (42). The use of alcohols in the mobile phase may also prevent hydrogen bond formation (43). Other examples of polymer adsorption onto SEC packings are given in references 44–60.

Electrostatic Interactions. Ion exchange and ion exclusion of polyelectrolytes are caused by the presence of dissociated silanol groups on silica-based packings, carboxylic groups on polymeric packings, or other ionized groups that have been introduced onto the packing. Because most chromatographic packings have surface anionic groups, which act as cationic exchange sites, electrostatic interactions are usually encountered. Cationic polyelectrolytes are adsorbed by ion exchange, and anionic polyelectrolytes are excluded from entering the pores of the packing because of electrostatic repulsive forces. These effects can be, in many cases, eliminated by adding electrolyte to the mobile phase, which helps to screen electrostatic forces. Also, ion exchange and exclusion can be reduced or prevented by lowering the pH (usually below 4) to suppress dissociation of silanol and carboxylic groups. In addition, a cationic compound, in some instances the sample itself, may be added to the mobile phase to deactivate anionic sites.

For the SEC analysis of polyelectrolytes, it is best to use a hydrophilic, polymeric packing or a surface-modified silica packing in which most of the silanols are derivatized. Another approach for the analysis of cationic polyelectrolytes is to use packings derivatized with cationic functional groups (61). This is the basis of a commercially available SEC packing which is chemically bonded with polyethylenimine (62).

A dramatic example of ion exclusion of a water-soluble polymer, which at first was thought to be "nonionic", is shown in Figure 2. When analyzed using an aqueous mobile phase (methanol was added to prevent hydrophobic interaction between polymer and packing), an unusual multimodal distribution was obtained indicating unexpected high molecular weight components. With the addition of 0.05 M $LiNO_3$ to the mobile phase, a typical peak shape was obtained. (The peak eluting at 34 cm was caused by mobile phase mismatch and ion inclusion.) Subsequent analysis by

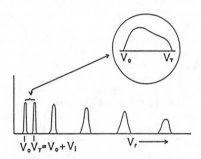

Figure 1.   The relationship between SEC and other chromatographic techniques. V_0 is the elution volume of an excluded peak (interstitial volume), V_T is the total permeated peak volume $(V_0 + V_i)$, and V_r is the retention volume of an adsorbed component.   Reproduced with permission from reference 8.   Copyright 1984, Astor Publishing Corp.

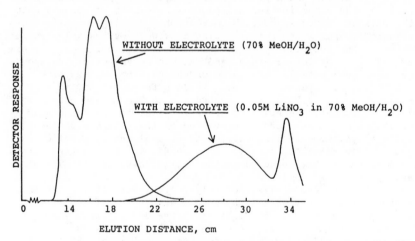

Figure 2.   An example of ion exclusion of a water-soluble polymer lightly substituted with carboxylic groups.

titrimetry showed that this "nonionic" water-soluble polymer was lightly substituted with carboxylic groups, which were inadvertently introduced during synthesis.

Ion inclusion, which was first reported by Stenlund (63) and investigated by others (64-67), is a rather unusual "electrostatic" nonsize-exclusion mechanism that occurs when analyzing polyelectrolytes. Ion inclusion is not really an enthalpic interaction, but a perturbation of the chemical potential between ionic components in the interstitial volume and those in the pore volume. During elution, polyelectrolyte counterions have the potential of freely diffusing into the pores of the packing ($K_d = 1$) as compared to the polyelectrolyte which is size excluded ($K_d < 1$). Because electroneutrality must be established between species in the pores and those in the interstitial volume, additional polymer is forced into (ion included) the packing to relax the chemical potential difference. This effect is caused by the establishment of Donnan membrane equilibrium. As a result, polyelectrolytes may have larger K_d values than expected. Ion inclusion of polyelectrolytes can be readily eliminated by adding electrolyte to the mobile phase; however, this leads to the presence of a total permeation "salt" peak caused by the inclusion of the electrolyte ion having the same charge as the polyelectrolyte.

Recent studies on the SEC analysis of anionic polyelectrolytes are given in references 68-78, and those for cationic polyelectrolytes are covered by references 55,58,61,62, 79-91. Papers dealing with adsorption of proteins during SEC are 9,30, 92-102. The reader should also refer to reviews 16 and 17 for older references on these topics.

Intermolecular Electrostatic Effects

Because of fixed charges on polyelectrolytes, they exhibit highly unusual solution properties (103). In low ionic strength solvents, electrostatic repulsive forces among neighboring ionic sites expand the polymer chain, increasing its hydrodynamic volume. With the addition of electrolyte to the solvent, these electrostatic forces are screened and the polymer contracts. The "salt-sensitivity" of polyelectrolytes can be quite dramatic. For example, in the case of carboxymethylcellulose, the intrinsic viscosity decreased from 80 to 4.6 dL/g when the ionic strength was increased from close to zero to 0.7M (104). The extent of polyelectrolyte contraction is governed by the degree of dissociation and ionic substitution on the polyelectrolyte, as well as by the uniformity and location of the ionic groups on the polymer chain.

Polyelectrolyte contraction can be followed by determining the K_d as a function of mobile phase ionic strength (104). In practice, however, the mobile phase ionic strength must be sufficiently high to ensure that the chain is in a contracted state. In this way, small changes in ionic strength, which may be inadvertently introduced during mobile phase preparation, will not affect the elution behavior of the sample. Also, if the ionic

strength of an injected sample solution is different from that of
the mobile phase, the mobile phase ionic strength should be
sufficiently high to off-set any difference. See references
63,65,66,105-110 for examples.

Another peculiar property of polyelectrolytes that takes
place in low ionic strength solutions is the expansion of the chain
as the polymer concentration is decreased. This phenomenon is
caused by a decrease in concentration of closely associated
counterions surrounding the polyelectrolyte as the polymer
concentration is decreased (111). This effect results in decreased
electrostatic screening among ionic sites on the polymer, leading
to polymer expansion. Because of fixed ionic charges on the
polymer, intramolecular osmotic pressure also causes molecular
expansion. Thus, if polyelectrolytes are analyzed by using
relatively low ionic strength mobile phases, severe peak fronting
results (112). Because the polymer concentration is lower on
either side of the peak maximum, the polymer is expanded in these
regions and elutes at a higher velocity (has a smaller K_d value)
than the peak maximum; a distorted peak profile results. With
added electrolyte, the intramolecular electrostatic repulsive and
osmotic forces are reduced.

Concentration Effects

For high molecular weight polymers, concentration effects are
manifested by increased elution volumes and, if the viscosity of
the injected solution is significantly higher than that of the
mobile phase, by peak distortion. These effects can be explained
in terms of two phenomena: macromolecular crowding and viscous
fingering. At high polymer concentrations, a critical
concentration is approached whereby segmental chain motion becomes
somewhat restricted because of chain overlap. This reduces the
hydrodynamic volume of the polymer caused by volume constraints
imposed by neighboring polymers. In addition, macromolecular
crowding will decrease the conformational entropy of the polymer in
the interstitial volume, thereby increasing the $\Delta S°$ term (making
it less negative) in eqn. 3, leading to an increase in K_d. Peak
distortion will result if the viscosity difference between the
mobile phase and injected sample is sufficiently large enough to
cause perturbation of the velocity streamlines of the eluting
sample. Thus, the mobile phase "fingers" its way through the
solute plug, resulting in severely distorted peaks (113). If the
injected concentration exceeds the pore volume capacity, normal
chromatographic overloading will occur. However, in this case, the
polymer elution volume will decrease. Theoretical models that take
into account these three concentration effects have been proposed
by Janca (114-125).

A major problem with viscous fingering is that reproducible
peak shapes, albeit distorted, may be observed; this may be highly
misleading in interpreting SEC results. To obtain reliable
results, molecular weight distributions of samples should be
obtained as a function of polymer concentrations to arrive at a
value in which peak distortion is not present or peak shape does

not change with further lowering of concentration. Different approaches used to correct for concentration effects are found in references 126-133 and reviewed in 28 and 134.

It is rather difficult to establish guidelines concerning recommended relative viscosities to be used to prevent concentration effects. The major reason for this is that viscous drag not only depends on the rheological characteristics of the polymer solution (115,119), but also on the amount of longitudinal and lateral dispersion (i.e., dilution) a sample undergoes as it travels through the column. Also, the mode of injection (e.g., point source or diffused sample entry into the column) will also play a role regarding chromatographic viscosity effects. For example, in conventional aqueous SEC, it has been recommended that the relative viscosity of the sample as compared to the mobile phase should be less than 2 (135). However, in a high performance column where there is less peak dispersion, this value will be considerably lower (104,105,114-120,136).

Additional studies on SEC concentration effects can be found in references 76,137-147.

Polymer Shear Degradation

Until recently, there has been, surprisingly, very little data on shear degradation of polymers during SEC (12, 148-151). Shear degradation in a packed bed is a rather complex hydrodynamic process and depends on a number of parameters including shear rate, elongational strain rate, polymer concentration, the nature of the solvent, and the chemical structure of the polymer. In addition to the packed bed, other parts of the chromatographic system with which the polymer comes into contact may generate sufficient shear forces to cause polymer degradation. Injector valve passages and sample loops, column frits, or capillary tubing used for connections and within the detector may be possible sites of shear degradation. Compared to conventional SEC, linear velocities generated in high performance SEC columns are typically five to ten times greater. This combined with the use of much smaller packing particle sizes lead to shear rates that are one to two orders of magnitude higher than for conventional columns. From a survey of literature values of critical molecular weights and shear rates (148), and the onset of shear degradation observed in SEC, it appears that shear rates in SEC systems must be kept below 10^3-10^4 sec^{-1} to avoid degradation. In most reported studies, the molecular weight above which shear degradation was observed was ~5×10^5. For the analysis of ultrahigh molecular weight polymers (>10^6 g/mole), flow rates of less than 0.1 mL/min may be required when using a 4mm ID column. Because of the inverse relationship between particle diameter and shear rate, the use of SEC packings much smaller than 5 μm should be avoided for the analysis of >10^6 molecular weight polymers.

Experimentally it is difficult to detect the occurrence of polymer shear degradation since concentration effects, increased peak dispersion, and ultrafiltration of high molecular weight components may also distort the peak profile or shift the distribution towards the low molecular weight region. Furthermore,

because shear degradation may occur anywhere in the chromatographic
system where a high force field exists, shear degradation may not
necessarily be accompanied by an increase in elution volume or
tailing, especially if degradation occurs near the end of or after
the SEC column.

A qualitative approach for detecting shear degradation is to
examine the shape of a calibration curve generated by the use of a
"molecular weight" detector (low-angle laser light scattering
photometer or viscosity detector): A downward shift of the log M
versus V_e plot is indicative of shear degradation. The best
approach to determine the extent of shear degradation is to measure
either the intrinsic viscosity or an average molecular weight of
the polymer before and after elution through the column.

Ultrafiltration

With decreasing packing size in SEC columns, the probability of
physical entrapment of macromolecules increases. To estimate the
molecular weight limit above which ultrafiltration will occur, we
must first calculate an average radius of the interstices formed in
a packed bed. This is done by assuming that the packed column
consists of a bundle of capillaries in which the capillary radius
can be estimated from the bed hydraulic radius:

$$R_h = D_p \varepsilon / 6(1-\varepsilon) \tag{7}$$

where D_p is the average diameter of the packing and ε is the
porosity of a packed bed taken as 0.36. Thus,

$$R_h = 0.094 \, D_p \tag{8}$$

It should be noted that if there are any fines present, which would
readily fill in the interstices, R_h in those regions of the
packed bed would be considerably smaller.

In addition to the packed bed acting as an ultrafilter, the
porous frits used at both ends of the column may act as very
effective filtering devices. Thus a 2-μm porosity frit would
have an average pore radius of 1 μm. Because of the tortuosity
and relatively wide pore-size distribution present in frits, it
would be safe to assume that it contains much smaller crevices
which can entrap macromolecules.

The molecular weight of a polymer which begins to approach
R_h can be approximated by calculating the radius-of-gyration of a
macromolecule $<s^2>^{1/2}$, which is defined as the
root-mean-square distance of the elements of the chain from its
center of gravity, using the Flory-Fox equation (111):

$$<s^2>^{3/2} = [\eta]M/\Phi' \tag{9}$$

where [η] is the intrinsic viscosity, M is the molecular weight,
and Φ' is a constant that is equal to 0.39×10^{25} mol^{-1} when
[η] is expressed in terms of cm^3/g. A more general equation
written in terms of the Mark-Houwink equation is

$$\langle s^2 \rangle^{3/2} = KM^{a+1}/\Phi \qquad (10)$$

where K and a are the Mark-Houwink constants.
The molecular weight equivalents to R_h for different
packing sizes for several representative polymers, calculated from
eqn. (9), are shown in Table I. As indicated, the use of 5- and
10- μm packings should pose no serious problem because most
polymers commonly encountered have molecular weights below
1 x 10^7 g/mol. The use of $\leq$ 2 μm packings, however, may lead
to ultrafiltration depending upon the polydispersity of the
sample. These conclusions are based on the assumption that there
are no fines present in the column. When analyzing ultrahigh
molecular weight polymers using high porosity (>3000 Å pore
diameters) silica packings, this assumption is probably not true
because of the fragility of these packings (155). Thus, the
presence of, let us say, 1-μm fragments caught in the interstices
of 10 μm packings could result in <0.1 μm voids which are
sufficiently small to trap, for example, >2 x 10^6 g/mol
polyethylene. Furthermore, insoluble material present in the
sample may also become trapped either in the frit or the column
packing and further reduce the interstices.

Table I. Molecular Weight Equivalent to the Hydraulic Radius
of a Column Packed with 2-, 5-, and 10-μm Particles

	D_p, μm		
Polymer	2	5	10
Polystyrene[a]	1.4x10^7	6.6x10^7	2x10^8
Linear Polyethylene[b]	7.1x10^6	3.5x10^7	1.2x10^8
Poly(styrene sulfonate)[c]	7.2x10^6	3.1x10^7	9.3x10^7

a. a = 0.766, K = 6.82 x 10^{-3} in THF (152).
b. a = 0.7, K = 5.9x10^{-2} in trichlorobenzene (153).
c. a = 0.89, K = 2.8x10^{-3} in 0.01M NaCl (154)

Hydrodynamic Effects

Hydrodynamic or flow-rate effects in SEC have been reviewed
recently by Aubert and Tirrell (156) and Giddings (12). Although
Aubert and Tirrell propose that nonequilibrium effects are not
significant in SEC, they show experimental evidence to support an
effect called molecular migration (157) in which K_d increases
with flow rate. One possible mechanism that is proposed is that
nonhomogeneous and curvilinear flow fields, which exist in porous
media flow, cause macromolecules to migrate to the packing surface
(concave side of streamlines). This enhanced concentration forces
more solute into the packing, thus increasing K_d.

Giddings ($\underline{12}$) discusses the consequences of finite flow rate within the pores of large porosity packings, which are required for the analysis of ultrahigh molecular polymers. Another possible phenomenon that is presented is concentration polarization. This is caused by pore flow in which partially excluded polymer concentrates on the surface of the packing, thereby increasing K_d. If indeed present, polymer at the surface of the pore may also form a secondary exclusion barrier which would possibly act to exclude smaller macromolecules, decreasing K_d of lower molecular weight components.

Counter to the above phenomena is an effect called hydrodynamic chromatography in which high molecular weight macromolecules under laminar flow conditions tend to align themselves away from surfaces. The reason for this is that large macromolecules cannot approach surfaces because of steric exclusion. As a result, they do not sample the low velocity streamlines, which are present near the surface, but are focused radially towards higher velocity streamlines. In the hydrodynamic chromatographic separation mode, which is done in either capillary tubes or columns packed with nonporous packings, the larger particle or macromolecule elutes first. Although this technique has been used for particle size determinations ($\underline{158}$), it has been recently applied to macromolecules ($\underline{159-163}$). If this effect is superimposed on the steric exclusion mechanism in SEC, it would favor the SEC separation.

Polymer Chain Orientation and Deformation

It has been reported that elongational and shear flow fields can orient macromolecules as well as change their conformation ($\underline{159-162,164,165}$). Although Aubert and Tirrell ($\underline{156}$) suggest that this effect is not significant in SEC, Prud'homme and coworkers ($\underline{159-162}$) have explained the elution behavior of high molecular weight polymers observed in hydrodynamic chromatography on the basis of chain orientation and deformation. Deformation of the random-coil shape occurs when the Deborah number, which is defined as the longitudinal velocity gradient times the relaxation time of a molecule in solution, is greater than 0.5. In the recent study by Langhorst, et al. ($\underline{162}$) a flow rate of approximately 0.025 mL/min through a 10-mm ID column packed with 15-μm cation-exchange resin, was required to keep the Deborah number below 0.5.

If polymer orientation and deformation were to occur in SEC, which is possible because of the extremely high shear rates encountered in high performance systems ($\underline{148}$), K_d values would be smaller than expected and would be dependent on flow rate. In SEC however, unlike hydrodynamic chromatography, there is a partitioning process in which the polymer diffuses from the high flow fields, through a quiescent layer of mobile phase on the surface of the packing, and into the almost stagnant liquid phase within the pores of the packing. Along this path, elongational force fields are extremely small, except for the boundary between the high velocity streamlines in the mobile phase and at the surface or the entrance of pores. As a result, macromolecules would not be subjected to orientational or elongational force fields during partitioning, which is in agreement with Aubert and Tirrell's remarks (156).

Summary

As in all other liquid chromatographic techniques, extreme care must be taken in choosing the proper mobile phase/packing combination to ensure adequate resolution and the absence of mixed mechanisms of separation. In a typical SEC experiment it may be difficult, if not impossible, to determine which nonsize-exclusion effect is occurring. This is especially true when analyzing ultrahigh molecular weight polymers in which there could be a number of effects occurring simultaneously. Because SEC is a relative technique that depends on calibration with known standards, nonideal behavior may at times be factored out. Criteria that may be used in judging whether or not the separation is based strictly on steric exclusion include measuring elution as a function of sample concentration, temperature, and flow rate. Also, in the absence of secondary separation mechanisms, the universal calibration should hold.

Nonsize-exclusion effects are very interesting per se and offer an opportunity of studying dilute solution properties of polymers such as electroviscous effects, macromolecular crowding, and shear degradation. Enthalpic integrations, although detrimental for SEC, provide a powerful separation technique for determining molecular weight distributions and chemical hetereogeneity of polymers.

Since we are now in the era of high performance SEC, in which column efficiency and speed of analysis are maximized, limitations of SEC mainly for the analysis of high molecular weight polymers are becoming more apparent. There are a number of secondary separation mechanisms that may come into play, which were not evident with conventional columns, because of the high force fields generated in high performance columns. Nevertheless, SEC is still the premier technique for rapidly obtaining the molecular weight distribution of polymers covering a molecular weight range of over six-orders of magnitude.

Literature Cited

1.   Cassasa, E. F.; Tagami, Y. *Macromolecules* 1969, *2*, 14.
2.   Mori, S.; Suzuki, T. *Anal. Chem.* 1980, *52*, 1625.
3.   Dawkins, J. V.; Hemming, M. *Makromol. Chem.* 1975, *176*, 1795.
4.   Dawkins, J. V. *J. Polym. Sci.*, *Polym. Phys. Ed.* 1976, *14*, 569.
5.   Dawkins, J. V. *J. Chromatogr.* 1977, *135*, 470.
6.   Dawkins, J. V. *Polymer* 1978, *19*, 705.
7.   Dawkins, J. V. *Pure Appl. Chem.* 1979, *51*, 1473.

8.   Barth, H. G. LC Mag. 1984, 2, 24.
9.   Kopaciewicz, W.; Regnier, F. E. Anal. Biochem. 1982, 126, 8.
10. Mori, S.; Yamakawa, A. Anal. Chem. 1979, 51, 382.
11. Campos, A.; Borque, L.; Figueruelo, J. E. J. Chromatogr.
     1977, 140, 219.
12. Giddings, J. C. Adv. Chromatogr. 1982, 20, 217.
13. Yau, W. W.; Kirkland, J. J.; Bly, D. D. "Modern Size
     Exclusion Liquid Chromatography"; Wiley: New York, 1979.
14. Janca, J., Ed. "Steric Exclusion Liquid Chromatography of
     Polymers"; Dekker:New York, 1984.
15. Audebert, R. Polymer 1979, 20, 1561.
16. Barth, H. G. J. Chromatogr. Sci. 1980, 18, 409.
17. Barth, H. G. Adv. Chem. Ser. 1985, 213, 31.
18a. Hagnauer, G. L. Anal. Chem. 1982, 54, 265R.
18b. Majors, R. E.; Barth, H. G.; Lochmuller, C. H. Anal. Chem.
     1984, 56, 300R.
19. Barth, H. G.; Barber, W. E.; Lochmuller, C. H.; Majors, R.
     E.; Regnier, F. E. Anal. Chem. 1986, 58, 211R.
20. Karger, B. L.; Snyder, L. R.; Horvath, C. "An Introduction
     to Separation Science"; Wiley:New York, 1973.
21. Dawkins, J. V.; Hemming, M. Polymer 1975, 16, 554.
22. Dawkins, J. V. J. Liq. Chromatogr. 1978, 1, 279.
23. Mori, S. Anal. Chem. 1978, 50, 745.
24. Bakos, D.; Bleha, T.; Ozima, A.; Berek, D. J. Appl. Polym.
     Sci. 1979, 23, 2233.
25. Belenkii, B. G.; Vilenchik, L. Z.; Nesterov, V. V.; Kolegov,
     V. J.; Frenkel, S. Ya J. Chromatogr. 1975, 109, 233.
26. Lecourtier, J.; Audebert, R.; Quivoron, C. J. Chromatogr.
     1976, 121, 173.
27. Lecourtier, J.; Audebert, R.; Quivoron, C. Pure Appl. Chem.
     1979, 51, 1983.
28. Mori, S. In "Steric Exclusion Chromatography of Polymers";
     Janca, J., Ed.; Dekker:New York, 1984; Vol. 25, p. 161.
29. Belenkii, B. G.; Vilenchik, L. Z. "Modern Liquid
     Chromatography of Macromolecules"; Dekker: New York, 1984.
30. Hefti, F., Anal. Biochem. 1982 121, 378.
31. Murray, G. J.; Youle, R. J.; Gandy, S. E.; Zirzow, G. C.;
     Barranger, J. A. Anal. Biochem. 1985, 147, 301.
32. Vivilecchia, R. V.; Lightbody, B. G.; Thimot, N. Z.; Quinn,
     H. M. J. Chromatogr. Sci. 1977, 15, 424.
33. Barth, H. G. Anal. Biochem 1982, 124, 191.
34. Schmidt, D. E.; Giese, R. W.; Connor, D.; Karger, B. L.
     Anal. Chem. 1980, 52, 177.
35. Pfannkoch, E.; Lu, K. C.; Regnier, F. E.; Barth, H. G. J.
     Chromatogr. Sci. 1980, 18, 430.
36. Haglund, A. C.; Marsden, N. V. B. J. Polym. Sci., Polym.
     Lett. Ed. 1980, 18, 271.
37. Marsden, N. V. B.; Haglund, A. C. J. Inclusion Phenomena
     1984, 2, 21.
38. Haglund, A. C.; Marsden, N. V. B. J. Chromatogr. 1984, 301,
     47.
39. Haglund, A. C.; Marsden, N. V. B. J. Chromatogr. 1984, 301,
     365.

40. Haglund, A. C.; Marsden, N. V. B.; Ostling, S. G. J. Chromatogr. 1985, 318, 57.
41. Radhakrishnamurthy, N.; Jeansonne, N.; Berenson, G. S. J. Chromatogr. 1983, 256, 341.
42. Stone, R. G.; Krasowski, J. A. Anal. Chem. 1981, 53, 736.
43. Figueruelo, J. E.; Soria, V.; Campos, A. Makromol. Chem. 1981, 182, 1525.
44. Adams, J. R.; Bicking, M. K. L. Anal. Chem. 1985, 57, 2844.
45. Berek, D.; Bakos, D.; Bleha, T.; Soltes, L. Makromol. Chem. 1975, 176, 391.
46. Biran, R.; Dawkins, J. V. Eur. Polym. J. 1984, 20, 129.
47. Campos, A.; Soria, V.; Figueruelo, J. E. Makromol. Chem. 1979, 180, 1961.
48. Campos, A.; Figueruelo, J. E. Makromol. Chem. 1977, 178, 3249.
49. Cesteros, C.; Katime, I.; Strazielle, C. Makromol. Chem., Rapid Commun. 1983 4, 193.
50. Chiantore, O. J. Liq. Chromatogr. 1984, 7, 1.
51. Chiantore, O. J. Liq. Chromatogr. 1984, 7, 1867,
52. Dawidowioz, A. L.; Sokolowski, S. Chromatographia 1984, 18, 579.
53. Dawidowicz, A. L.; Sokolowski, S. Z. Phys. Chem. 1984, 265, 526.
54. Ludlam, P. R.; King, J. G. J. Appl. Polym. Sci. 1984, 29, 3863.
55. Mencer, H. J.; Grubisic-Gallot, Z. J. Chromatogr. 1982, 241, 213.
56. Misra, N.; Mandal, B. M. Makromol. Chem., Rapid Commun. 1984, 5, 471.
57. Ogawa, T.; Sakai, M. J. Liq. Chromatogr. 1983, 6, 1385.
58. Rand, W. G.; Mukherji, A. K. J. Chromatogr. Sci. 1982, 20, 182.
59. Saitoh, K.; Ozaki, E.; Suzuki, N. J. Chromatogr. 1982, 239, 661.
60. Spychaj, T.; Berek, D. Acta. Polym. 1982, 33, 479.
61. Talley, C. P.; Bowman, L. M. Anal. Chem. 1979, 51, 2239.
62. Gooding, D. L.; Schmuck, M. N.; Gooding, K. M. J. Liq. Chromatogr. 1982, 5, 2259.
63. Stenlund, B. Adv. Chromatogr. 1976, 14, 37.
64. Lindstrom, T.; deRuvo, A.; Soremark, C. J. Polym. Sci., Polym. Chem. Ed. 1977, 15, 2029.
65. Buytenhuys, F. A.; van der Maeden, F. P. B. J. Chromatogr. 1978, 149, 489.
66. Domand, A.; Rinaudo, M.; Rochas, C. J. Polym. Sci., Polym. Phys. Ed. 1979, 17, 673.
67. Rochas, C.; Nomard, A.; Rinaudo, M. Eur. Polym. J. 1980, 16, 135.
68. Bose, A.; Rollings, J. E.; Caruthers, J. M.; Okos, M. R.; Tsao, G. T. J. Appl. Polym. Sci. 1982, 27, 795.
69. Deckers, H. A.; Olieman, C.; Rombouts, F. M.; Pilnik, W. Carbohydr. Polym. 1986. 6, 361.
70. Djordjevic, S. V.; Batley, M.; Redmond, J. W. J. Chromatogr. 1986, 354, 507.

71.   Dubin, P. L.; Tecklenburg, M. M. Anal. Chem. 1985, 57, 275.
72.   Hann, N. D. J. Polym. Sci., Polym. Chem. Ed. 1977, 15, 1331.
73.   Kadkokura, S.; Miyamoto, T.; Inagaki, H. Makromol. Chem.
      1983, 184, 2593.
74.   Klein, J.; Westenkamp, A. J. Appl. Polym. Sci., Polym.
      Chem. Ed. 1981, 19, 707.
75.   Miyajima, T.; Yamauchi, K.; Ohashi, S. J. Liq. Chromatogr.
      1982, 5, 265.
76.   Rinaudo, M.; Desbrieres, J.; Rochas, C. J. Liq. Chromatogr.
      1981, 4, 1297.
77.   Scheuing, D. R. J. Appl. Polym. Sci. 1984, 29, 2819.
78.   Waki, H.; Tsuruta, K.; Tokunaga, Y. J. Liq. Chromatogr.
      1985, 8, 2105.
79.   Domard, A.; Rinaudo, M. Polym. Commun. 1984, 25, 55.
80.   Dubin, P. L. Sep. Purif. Methods 1981, 10, 287.
81.   Dubin, P. L.; Levy, I. J. J. Chromatogr. 1982, 235, 377.
82.   Dubin, P. L.; Levy, I. J.; Oteri, R. J. Chromatogr. Sci.
      1984, 22, 432.
83.   Guise, G. B.; Smith, G. C. J. Chromatogr. 1982, 235, 365.
84.   Guise, G. B.; Smith, G. C. J. Appl. Polym. Sci. 1985, 30,
      4099.
85.   Kato, Y.; Hashimoto, T. J. Chromatogr. 1982, 235, 539.
86.   Levy, I. J.; Dubin, P. L. Ind. Eng. Chem. Prod. Res. Dev.
      1982, 21, 59.
87.   Melawer, E. G.; DeVasto, J. K.; Frankoski, S. P.; Montana,
      A. J. J. Liq. Chromatogr. 1984, 7, 441.
88.   Mencer, H. J.; Vajnaht, Z. J. Chromatogr. 1982, 241, 205.
89.   Mori, S. Anal. Chem. 1983, 55, 2414.
90.   Stickler, M. Angew. Makromol. Chem. 1984, 123/124, 85.
91.   Stickler, M.; Eisenbeiss, F. Eur. Polym. J. 1984, 20, 849.
92.   Chang, J. P. J. Chromatogr. 1984, 317, 157.
93.   DeLigny, C. L.; Gelsema, W. J.; Roozen, A. M. P.
      J. Chromatogr. Sci. 1983, 21, 174.
94.   Engelhardt, H.; Ahr, G.; Hearn, M. T. W. J. Liq.
      Chromatogr. 1981, 4, 1361.
95.   Frigon, R. P.; Leypoldt, J. K.; Uyeji, S.; Henderson, L. W.
      Anal. Chem. 1983, 55, 1349.
96.   Galkin, A. V.; Kovaleva, I. E.; Kal'nov, S. L.
      Anal. Biochem. 1984, 142, 252.
97.   Kato, Y.; Hashimoto, T. J. High Resolut. Chromatogr. 1983,
      6, 45.
98.   Kato, Y.; Hashimoto, T. J. High Resolut. Chromatogr. 1983,
      6, 324.
99.   Kato, Y.; Hashimoto, T. J. High Resolut. Chromatogr. 1985,
      8, 78.
100. Link, G. W.; Keller, P. L.; Stout, R. W.; Banes, A. J.
      J. Chromatogr. 1985, 331, 253.
101. Roumeliotis, P.; Unger, K. K. J. Chromatogr. 1981, 218, 535.
102. Swergold, G. D.; Rubin, C. S. Anal. Biochem. 1983, 131, 295.
103. Mandel, M. Angew. Makromol. Chem. 1984, 123/124, 63.
104. Barth, H. G.; Regnier, F. E. J. Chromatogr. 1980, 192, 275.
105. Barth, H. G.; J. Liq. Chromatogr. 1980, 3, 1481.
106. Cha, C. Y. J. Polym. Sci., Polym. Lett. 1969, 7, 343.

107. Martenson, R. E. *J. Biol. Chem.* 1978, 253, 8887.
108. Cooper, A. F.; Matzinger, D. P. *J. Appl. Polym. Sci.* 1979, 23, 419.
109. Skalka, M. *J. Chromatogr.* 1968, 35, 456.
110. Spatorica, A. L.; Beyer, G. L. *J. Appl. Polym. Sci.* 1975, 19, 2938.
111. Flory, P. J. "Principles of Polymer Chemistry"; Cornell University: Ithaca, 1953.
112. Nefedov, P. P.; Lazareva, M. A.; Belenski, B. G.; Frenkel, S. Ya; Morton, M. M. *J. Chromatogr.* 1979, 170, 11.
113. Moore, J. C. *Sep. Sci.* 1970, 5, 723.
114. Janca, J. *J. Chromatogr.* 1977, 134, 263.
115. Janca, J.; Pokorny, S. *J. Chromatogr.* 1978, 148, 31.
116. Janca, J.; Pokorny, S. *J. Chromatogr.* 1978, 156, 27.
117 Janca, J. *J. Chromatogr.* 1979, 170, 309.
118. Janca, J.; Pokorny, S. *J. Chromatogr.* 1979, 170, 319.
119. Janca, J. *J. Chromatogr.* 1980, 187, 21.
120. Janca, J. *Anal. Chem.* 1979, 51, 637.
121. Janca, J. *Polym. J.* 1980, 12, 405.
122. Janca, J.; Pokorny, S.; Bleha, M.; Chiantore, O. *J. Liq. Chromatogr.* 1980, 3, 935.
123. Janca, J.; Pokorny, S.; Vilenchik, L. Z.; Belenkii, B. G. *J. Chromatogr.* 1981, 211, 39.
124. Janca, J. *J. Liq. Chromatogr.* 1984, 7, 1887.
125. Janca, J. *J. Liq. Chromatogr.* 1984, 7, 1903.
126. Cantow, M. J. R.; Porter, R. S.; Johnson, J. F. *J. Polym. Sci.* B 1966, 707, 4.
127. Boni, K. A.; Sliemers, F. A.; Stickney, P. B. *J. Polym. Sci.* A2, 1968, 6, 1567.
128. Goetz, K. P.; Porter, R. S.; Johnson, J. F. *J. Polym. Sci.*, A2, 1971, 9, 2255.
129. Rudin, A.; Wagner, R. A. *J. Appl. Polym. Sci.* 1976, 20, 1483.
130. Mori, S. *J. Appl. Polym. Sci.* 1976, 20, 2157.
131. Mori, S. *J. Appl. Polym. Sci.* 1977, 21, 1921.
132. Shi, L. H.; Ye, M. L.; Wang, W.; Ding, Y. K. *J. Liq. Chromatogr.* 1984, 7, 1851.
133. Linden, C. V. *Polymer* 1980, 21, 171.
134. Janca, J. *Adv. Chromatogr.* 1981, 19, 37.
135. "Gel Filtration – Theory and Practice"; Pharmacia:Uppsala, 1979.
136. Sandy, G.; Kubin, M. *J. Appl. Polym. Sci.* 1979, 23, 1879.
137. Bleha, T.; Bakos, D.; Berek, D. *Polymer* 1977, 18, 897.
138. Bleha, T.; Spaychaj, T.; Vondra, R.; Berek, D. *J. Polym. Sci.*, Polym. Phys. Ed. 1983, 21, 1903.
139. Chiantore, O.; Guaita, M. *J. Liq. Chromatogr.* 1985, 8, 1413.
140. Chiantore, O.; Guaita, M. *J. Liq. Chromatogr.* 1984, 7, 1867.
141. Figueruelo, J. E.; Campos, A.; Soria, V.; Tejero, R. *J. Liq. Chromatogr.* 1984, 7, 1061.
142. Holding, S. R.; Vlachogiannis, G.; Barker, P. E. *J. Chromatogr.* 1983, 261, 33.
143. Mingshi, S.; Guixian, H. *J. Liq. Chromatogr.* 1985, 8, 2543.
144. Rudin, A. *J. Polym. Sci.*, A1, 1971, 9, 2587.

145.   Soltes, L.; Berek, D.; Mikulasova, D.   Colloid Polym. Sci.
       1980, 258, 702.
146.   Soria, V.; Campos, A.; Tejero, R.; Figuereulo, J. E.;
       Abad, C.   J. Liq. Chromatogr. 1986, 9, 1105.
147.   Tejero, R.; Soria, V.; Campos, A.; Figueruelo, J. E.;
       Abad, C.   J. Liq. Chromatogr. 1986, 9, 711.
148.   Barth, H. G.; Carlin, F. J.   J. Liq. Chromatogr. 1984, 7,
       1717.
149.   Barth, H. G.   Proc. SPE 43rd Ann. Tech. Conf., 1985, 275.
150.   Rooney, J. G.; Verstrate, G.   In "Liquid Chromatography of
       Polymers and Related Materials III"; Vol. 19; Dekker:New
       York, 1981, p 207.
151.   McIntyre, D.; Shih, A. L.; Savoca, J.; Seeger, R.;
       MacArthur, A.   ACS Symp. Ser. 1984, 245, 227.
152.   Boni, K. A.; Sliemers, P. A.; Stickney, P. B.   J. Polym.
       Sci. A2 1968, 6, 1579.
153.   Ram, A.; Miltz, J.   J. Appl. Polym. Sci. 1972, 15, 2639.
154.   Takahashi, A.; Kato, T.; Nagasawa, M.   J. Phys. Chem. 1967,
       71, 2001.
155.   Knox, J. H.; McLennan, F.   J. Chromatogr. 1979, 185, 289.
156.   Aubert, J. H.; Tirrell, M.   J. Liq. Chromatogr. 1983,
       6(S-2), 219.
157.   Aubert, J. H.; Prager, S.; Tirrell, M.   J. Chem. Phys. 1980,
       73, 4103.
158.   Penlidis, A.; Hamielec, A. E.; MacGregor, J. F.   J. Liq.
       Chromatogr. 1983, 6(S-2), 179.
159.   Prud'homme, R. K.; Froiman, G.; Hoagland, D. A.   Carbohydr.
       Res. 1982, 106, 225.
160.   Prud'homme, R. K.; Hoagland, D. A.   Sep. Sci. Technolog.
       1983, 18, 121.
161.   Hoagland, D. A.; Larson, K. A.; Prud'homme, R. K.   In
       "Modern Methods of Particle Size Analysis"; Barth, H. G.,
       Ed.; Wiley:New York, 1984; Chpt. 9.
162.   Langhorst, M. A.; Stanley, F. W.; Cutie, S. S.; Sugarman,
       J. H.; Wilson, L. R.; Hoagland, D. A.; Prud'homme, R. K.
       Anal. Chem. 1986, 58, 2242.
163.   Lecourtier, J.; Chauveteau, G.   Macromolecules 1984, 17,
       1340.
164.   Durst, F.; Haas, R.; Kaczmar, B. U.   J. Appl. Polym. Sci.
       1981, 26, 3125.
165.   Haas, R.; Durst, F.   Rheol. Acta 1982, 21, 566.

RECEIVED April 7, 1987

Chapter 3

Preparative Gel Permeation Chromatography

Juris L. Ekmanis

Waters Chromatography Division, Millipore Corporation,
Milford, MA 01757

Preparative GPC using large Styragel columns (57mm I.D.
X 122 cm) has been used to fractionate polymer samples
in 50 min runs. A single 10^5Å Styragel GPC column was
used to fractionate a high molecular weight polystyrene
in order to obtain a wide range of different molecular
weight fractions for testing or for use as calibration
standards. Use of a dual column preparative system
(10^5, 10^3Å) improved the fractionation of the lower
molecular weight components of this polystyrene
sample. Similarly, a lower pore size column (one 10^3Å)
was used to fractionate a low molecular weight epoxy
resin.

Starting materials and purified fractions were
analyzed by GPC with appropriate sets of high efficiency Ultrastyragel columns to demonstrate the extent
of fractionation. The maximum loading capacity of the
preparative columns was predicted by first determining
the loading capacity on a corresponding analytical
column (7.8 mm I.D.) and then using an appropriate
scale factor to account for the differences in diameters of the columns. The maximum loading capacity of
high molecular weight materials is limited by sample
concentration effects while that of low molecular
weight materials (e.g.<5000) is limited by column
capacity.

The technique of Gel Permeation Chromatography (GPC) was introduced
by Moore and Hendrickson (1,2) in 1964 for determining molecular
weight distributions of polymer samples. The chromatographic column
packings used for this new technique consisted of porous spheres of
crosslinked styrene-divinyl benzene resins (37-75µm) that were subsequently available as a family of columns under the name Styragel.
Analytical column dimensions were 7.8 mm I.D. X 4 ft (122 cm).
Larger diameter columns were also available for preparative chromatography. In later years, GPC analysis times were reduced and
resolution was improved by using shorter columns that were packed
with smaller particle size material. A typical family of GPC
columns that is available today contains 7µm particles

0097-6156/87/0352-0047$06.00/0
© 1987 American Chemical Society

(Ultrastyragel; 7.8 mm I.D. X 30 cm) and affords four to five times the plates in one-fourth the analysis time as compared with the original Styragel columns.

Although the small particle size, high efficiency columns are used today for analytical GPC, the large particle size materials (Styragel) are still useful for larger scale preparative separations (3,4). Preparative GPC is useful for

o Preparation of calibration standards - narrow dispersity standards for GPC calibration.
o Polymer fractionation for subsequent analysis - ancillary techniques (IR, UV, etc.).
o Altering polymer dispersity - removing highest and/or lowest molecular weight fractions to obtain a sample with narrower dispersity (Mw/Mn).
o Isolation of additives.

In this paper, preparative GPC using large diameter Styragel columns (57 mm I.D. X 4 ft.) was used to fractionate polymer samples in 50 min on a single column. Two examples include the preparative separation of a high molecular weight polystyrene and a low molecular weight epoxy resin. A preparative separation of the high molecular weight polystyrene was also done using two columns to demonstrate the benefits of adding another column (lower pore size). These results demonstrate that the sample loading capacity is limited by concentration effects (viscosity, etc.) for high molecular weight polymers and by the loading capacity of the column for low molecular weight samples (<5000 mol. wt.) which do not exhibit concentration effects.

Experimental

The overall procedure was similar for each sample (high molecular weight polystyrene, low molecular weight epoxy resin) and is outlined in Table I.

Table I. Experimental Procedure

Procedure	Columns	High MW	Low MW	Flow Rate	Relative Load
Analytical (starting materials and fractions from prep runs).	Ultrastyragel	$10^5,10^4,10^3$Å	$10^3,500$Å	1 ml/min	-
Loading Study	Styragel	10^5Å (10^3Å)	10^3Å	1.25 ml/min	1
Preparative Fractionation	Styragel	10^5Å (10^3Å)	10^3Å	50 ml/min	40

In each case, the sample was first analyzed on an optimized set
of high efficiency Ultrastyragel columns. In order to use only one
column for the preparative separation, the highest pore size column
in the analytical set was selected so as not to exclude a significant amount of high molecular weight material. In the case of the
polystyrene sample, a preparative separation was also performed
using two columns to improve the fractionation of the lower molecular weight components.

Rather than use the preparative Styragel column directly, a
loading study was first performed on a narrow version (7.8 mm I.D. X
4 ft) of the preparative column in order to determine the optimum
sample load. Increasingly larger sample loads were then injected
onto the loading column with proportionately decreased detector
sensitivities until the maximum load was achieved. This was indicated by an increase in retention (concentration effects) or decrease in resolution and/or retention (column overloading). The
loading studies were performed at a flow rate of 1.25 ml/min. Since
the preparative column had a cross-sectional area 54 times as large
as that of the column in the loading study, the sample load and flow
rate (for constant linear velocity) could have been scaled up by
this factor and the same quality of separation would have been
achieved. In order to be somewhat conservative, a scale factor of
40 was used in this work. Thus, the flow rate used for the preparative separations was 50 ml/min.

Subsequent to the preparative separation, the original sample
and selected fractions were chromatographed on the analytical set of
high efficiency columns to demonstrate the degree of fractionation.

Tetrahydrofuran (THF, UV grade) was used as the mobile phase
throughout this work since the extent of fractionation could be
demonstrated by direct analysis of the preparative fractions without
the need for concentration. When samples are to be recovered by
removal of solvent, other mobile phases (methylene chloride, etc.)
may be preferred to avoid concentrating solvent impurities which are
formed in THF on exposure to air unless additional precautions are
taken.

Although a flow rate of 50 ml/min was used in these preparative
runs (typical column backpressure was 20 psi), it is acceptable to
use flow rates as high as 100 ml/min with the 57 mm I.D. Styragel
columns. Such increased flow rates would afford reduced analysis
times, especially when using several columns in series for improved
fractionation. Although the pumping system used for the preparative
work had a maximum flow rate of 80 ml/min, other systems are commercially available if higher flow rates are desired.

Analytical runs and loading studies were performed with columns
installed in conventional GPC systems (Waters) consisting of a Model
590 solvent delivery system, a U6K injector, an R401 differential
refractometer (or a Model 410 differential refractometer) and an
M730 Data Module. The preparative system consisted of a Model 590
solvent delivery system equipped with 80 ml/min pump heads (1/4"
plungers), an R403 differential refractometer and an M730 Data Module. Preparative sample solutions were loaded (manually) directly
through the solvent draw off valve of the M590 pump, eliminating the
need for an injection valve.

Results and Discussion

High Molecular Weight Polystyrene.
In the first example, a broad distribution sample of polystyrene was
analyzed on a set of high efficiency Ultrastyragel GPC columns
(10^5,10^4,10^3Å) with which it was determined that Mw = 214,000 and
Mn = 87,000. In the ideal case, a similar set of three (10^5,10^4,
10^3Å) preparative Styragel columns (each 57 mm I.D. X 4 ft) could
have been used to fractionate the polystyrene sample. However, we
initially chose to demonstrate the extent of fractionation possible
with only one preparative column and selected the 10^5Å column
packing material so as not to exclude any of the higher molecular
weight fractions.

The maximum loading capacity for this polystyrene sample was
determined using a flow rate of 1.25 ml/min on an analytical version
(7.8 mm I.D. X 4 ft) of the preparative Styragel column. With a
constant injection volume of 500 µl, sample concentration was increased and detector sensitivity proportionately decreased in Figure 1. Also indicated are the amounts of sample to be injected onto
the preparative column using a scale factor of 40. With increasing
sample size (mass) the chromatograms should remain unchanged if no
concentration effects (long elution) or column overloading effects
(shorter retention, loss of resolution) occur. No column overload
was observed but significant concentration effects (viscosity, etc.)
occurred at concentrations of ⩾3%. These concentration effects can
be reduced by decreasing sample concentration and increasing injection volume with constant sample load (Figure 2; A-D). Notice that
a further increase in sample load (Figure 2; E) revealed significant
increase in retention due to concentration effects. Similar concentration limiting effects also occurred with lower molecular
weight polymers, including polyvinyl chloride (Mw = 111,500; Mn =
53,300) and polycarbonate (Mw = 45,100; Mn = 19,500) where molecular
weight averages are based on narrow distribution polystyrene
standards.

The 500 µl injection of 2% polystyrene in THF (Figure 1) was
selected for scale up to preparative GPC on a 57 I.D. column. Using
a scale factor of 40, a 20 ml sample (2% polystyrene) containing
400 mg of polystyrene was injected and 50 ml fractions were collected (Figure 3). Molecular weight distributions of the starting
material and seven fractions (shaded areas in Figure 3) were determined on the analytical set of three Ultrastyragel columns
(Figure 4). Calculated molecular weight averages for these selected
fractions are listed in Table II. As expected, the best fractionation (narrowest distribution) was obtained in the higher molecular
weight region for which the 10^5Å columns is optimized. Improved
fractionation in the lower molecular weight region would require
substitution or addition of lower pore size columns.

In similar fashion, the 2ml injection of 1% polystyrene
(Figure2; D) was scaled up by a factor of 40. The preparative run
then consisted of an 800 mg load (80ml; 1% solution). Fractions
were collected as before and molecular weight averages were deter-

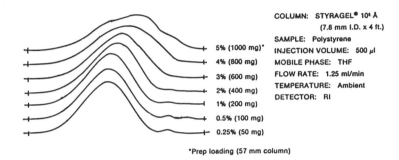

COLUMN: STYRAGEL® 10⁴ Å
(7.8 mm I.D. x 4 ft.)
SAMPLE: Polystyrene
INJECTION VOLUME: 500 μl
MOBILE PHASE: THF
FLOW RATE: 1.25 ml/min
TEMPERATURE: Ambient
DETECTOR: RI

5% (1000 mg)*
4% (800 mg)
3% (600 mg)
2% (400 mg)
1% (200 mg)
0.5% (100 mg)
0.25% (50 mg)

*Prep loading (57 mm column)

Figure 1. Polystyrene loading study on one column (Part 1).

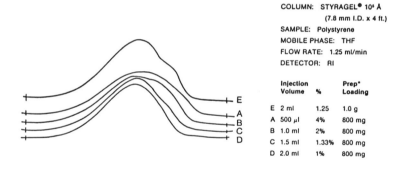

COLUMN: STYRAGEL® 10⁴ Å
(7.8 mm I.D. x 4 ft.)
SAMPLE: Polystyrene
MOBILE PHASE: THF
FLOW RATE: 1.25 ml/min
DETECTOR: RI

	Injection Volume	%	Prep* Loading
E	2 ml	1.25	1.0 g
A	500 μl	4%	800 mg
B	1.0 ml	2%	800 mg
C	1.5 ml	1.33%	800 mg
D	2.0 ml	1%	800 mg

*57 mm column

Figure 2. Polystyrene loading study on one column (Part 2).

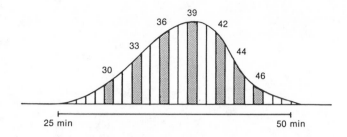

COLUMN: STYRAGEL® 10⁵ Å (57 mm I.D. x 4 ft.)
SAMPLE: Polystyrene, 2% (w/v)
INJECTION VOLUME: 20 ml (400 mg polymer)
MOBILE PHASE: THF
FLOW RATE: 50 ml/min
DETECTOR: RI (R403), 64X

Figure 3.   Preparative GPC of 400 mg polystyrene on one column.

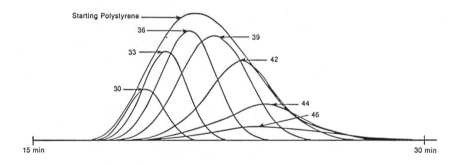

COLUMNS: ULTRASTYRAGEL™ 10⁵, 10⁴, 10³ Å
SAMPLE: Starting Polystyrene, preparative fractions
INJECTION VOLUME: 200 µl
MOBILE PHASE: THF
FLOW RATE: 1 ml/min
DETECTOR: RI

Figure 4.   GPC analysis of polystyrene fractions from
400 mg preparative run (one prep column).

Table II. Molecular Weight Averages of Polystyrene Fractions from
400 mg Preparative Run (10^5Å Column)

Fraction	Mw/10^3	Mn/10^3	Mw/Mn
30	543.3	471.4	1.15
33	365.7	302.0	1.21
36	237.5	179.4	1.32
39	150.1	100.4	1.50
42	91.6	54.3	1.69
44	68.5	32.1	2.14
46	90.2	21.9	4.13

mined. The results (Table III) indicated that the best fractionation again occurred in the highest molecular weight fractions and
that corresponding fractions exhibited slightly broader molecular
weight distributions in the 800 mg run than in the 400 mg run
(Table II).

Table III. Molecular Weight Averages of Polystyrene Fractions
from 800 mg Preparative Run (10^5Å Column)

Fraction	Mw/10^3	Mn/10^3	Mw/Mn
30	499.4	426.0	1.17
33	339.3	264.9	1.28
36	229.0	152.1	1.51
39	153.4	84.6	1.81
42	112.0	55.5	2.01
44	119.0	44.7	2.66
47	125.4	62.6	2.00

The separation of 400 mg polystyrene (Figure 3) was repeated
with two Styragel columns (Figures 5,6) to demonstrate the improved
fractionation of the lower molecular weight components that could be
achieved by the addition of a lower pore size column (10^3Å) to the
original 10^5Å column. Eight of the 50 ml fractions that were collected were subsequently analyzed on the analytical set of three
Ultrastyragel columns (Figure 7). Calculated molecular weight
averages for these selected fractions are listed in Table IV. A
comparison of corresponding fractions of the single and dual column
separations of 400 mg polystyrene (Table V) indicates that the
addition of a 10^3Å column to the single 10^5Å Styragel column
significantly improved the fractionation, especially for the lower
molecular weight fractions. As expected, the quality of the
fractions as judged by the narrowness (Mw/Mn) of the resulting
molecular weight distributions are very dependent on the column(s)
used for the separation.

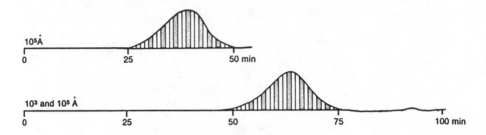

COLUMNS: Styragel® (each, 57 mm I.D. X 4 ft.)
SAMPLE: Polystyrene, 2% (w/v)
INJECTION VOLUME: 20 ml (400 mg polymer)
MOBILE PHASE: THF
FLOW RATE: 50 ml/min
DETECTOR: RI (R403), 64X

Figure 5.   Preparative GPC of 400 mg polystyrene (one vs. two column systems).

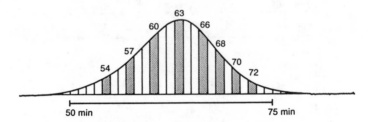

COLUMNS: Styragel® 10³, 10⁵ Å (each, 57 mm I.D. X 4 Ft.)
SAMPLE: Polystyrene, 2% (w/v)
INJECTION VOLUME: 20 ml (400 mg polymer)
MOBILE PHASE: THF
FLOW RATE: 50 ml/min
DETECTOR: RI (R403), 64X

Figure 6.   Preparative GPC of 400 mg polystyrene on two columns.

Table IV. Molecular Weight Averages of Polystyrene Fractions
from 400 mg Preparative Run (10^5, 10^3Å Columns).

Fraction	Mw/10^3	Mn/10^3	Mw/Mn
54	628.1	533.3	1.18
57	441.3	359.2	1.23
60	295.9	227.6	1.30
63	185.2	134.0	1.38
66	113.4	77.8	1.46
68	82.1	54.7	1.50
70	55.7	36.8	1.51
72	38.7	24.5	1.58

Table V. Comparison of Single Column and Dual Column
Preparative Separations of 400 mg Polystyrene

10^5Å Column		10^5 and 10^3Å Columns	
Fraction	Mw/Mn	Fraction	Mw/Mn
30	1.15	54	1.18
33	1.21	57	1.23
36	1.32	60	1.30
39	1.50	63	1.38
42	1.69	66	1.46
44	2.14	68	1.50
46	4.13	70	1.51
–	–	72	1.58

Low Molecular Weight Epoxy Resin.
A similar sequence was performed on a low molecular weight epoxy
resin to demonstrate that column capacity, not sample concentra-
tion/viscosity, determines the maximum sample load for low molecular
weight materials. The GPC analysis of a liquid epoxy on a set of
10^3 and 500Å Ultrastyragel columns as well as on a single 7.8 mm
I.D. X 4 ft 10^3Å Styragel column (for subsequent loading study) is
shown in Figure 8. The time scale in Figure 8 refers only to chromatogram A. The total volume of the column in chromatogram B is
50 ml and total run time is 40 min. The chart speed has been decreased from that in chromatogram A to facilitate a visual comparison of the relative resolution with these two systems. The loading
study (Figure 9) revealed no increase in elution volume (concen-
tration/viscosity effects) as was the case with the high molecular
weight samples. Only a slight decrease in resolution due to column
loading was observed at above 10% concentration (equivalent to 2.0 g
on a 57 mm preparative column). This eventual decrease in resolution was not due to concentration/viscosity effects since dilution
of the 12.5% solution by a factor of 4 and a proportional increase
in injection volume had no effect (Figure 9).

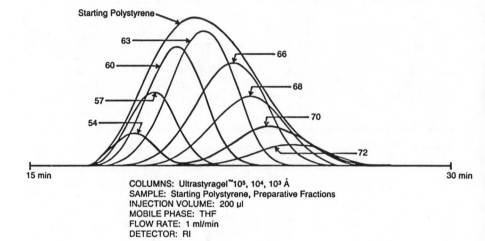

Figure 7. GPC analysis of polystyrene fractions from 400 mg preparative run (two prep columns).

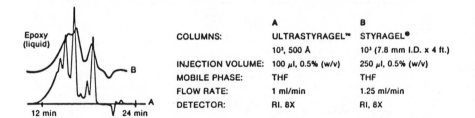

Figure 8. GPC analysis of low molecular weight epoxy resin.

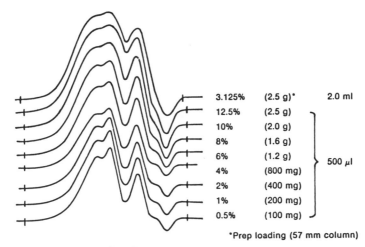

3.125%	(2.5 g)*	2.0 ml
12.5%	(2.5 g)	
10%	(2.0 g)	
8%	(1.6 g)	
6%	(1.2 g)	500 µl
4%	(800 mg)	
2%	(400 mg)	
1%	(200 mg)	
0.5%	(100 mg)	

*Prep loading (57 mm column)

COLUMN: STYRAGEL® 10³ Å (7.8 mm I.D. x 4 ft.)
SAMPLE: Epoxy
MOBILE PHASE: THF
FLOW RATE: 1.25 ml/min
DETECTOR: RI

Figure 9. Epoxy resin loading study on one column.

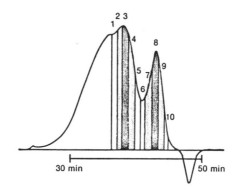

COLUMN: STYRAGEL® 10³ Å (57 mm I.D. x 4 ft.)
SAMPLE: Epoxy, 10% (w/v)
INJECTION VOLUME: 20 ml
MOBILE PHASE: THF
FLOW RATE: 50 ml/min
DETECTOR: RI (R403), 128X

Figure 10. Preparative GPC of 2.0 g epoxy resin on one column.

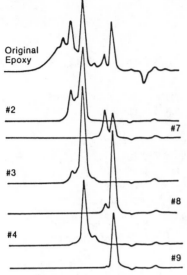

COLUMNS: ULTRASTYRAGEL™ 10³, 500 Å

INJECTION VOLUME: 100 µl

CONCENTRATION: Epoxy, 0.5%

Prep fractions, as collected

MOBILE PHASE: THF

FLOW RATE: 1 ml/min

DETECTOR: RI

Figure 11. GPC analysis of epoxy starting material and preparative fractions.

The 500 µl injection of 10% epoxy resin in THF (Figure 9) was selected for scale up to the preparative column (10³Å Styragel; 57 mm I.D. X 4 ft) using a scale factor of 40 for both injection volume and flow rate. Thus, a 20 ml injection (10% epoxy in THF) containing 2.0 g of epoxy resin was fractionated in Figure 10. Starting material and selected fractions (#2-4 and #7-9) were analyzed (Ultrastyragel columns) to demonstrate the extent of fractionation (Figure 11). If necessary, these initial fractions (e.g. #4,9) could be further purified on a set of high efficiency columns.

Summary

Preparative GPC with even a single column is useful for fractionation of both high and low molecular weight materials. Smaller diameter columns can be used to predict optimum sample loads for the preparative separations on large columns. Preparative loading is limited by concentration effects (viscosity, etc.) in the case of high molecular weight polymers and by column capacity for low molecular weight (<5000) materials where concentration/viscosity effects do not occur.

Literature Cited

1. Moore, J.C. J. Polym. Sci. 1964, A2, 835
2. Moore, J.C.; Hendrickson, J.G. J. Polym. Sci. 1965, C8, 233
3. Ekmanis, J.L. Pittsburgh Conference 1982, Paper No. 068
4. Ekmanis, J.L. Waters Journal of Analysis and Purification 1986, 1,75

RECEIVED May 28, 1987

Chapter 4

Orthogonal Chromatography and Related Advances in Liquid Chromatography

Stephen T. Balke

Department of Chemical Engineering and Applied Chemistry, University of Toronto, Toronto, Ontario M5S 1A4, Canada

Orthogonal chromatography (OC) and related recent advances in both fractionation and detection are reviewed. OC is a method of analyzing complex polymers using two interconnected size exclusion chromatographs to obtain a cross-fractionation. Conventional size exclusion chromatography (SEC) and high performance liquid chromatography (HPLC) analysis of complex polymers have focussed on detection and fractionation respectively. The development of OC as a method which combines the best of both approaches is summarized. In OC of linear copolymers, the fractionation is intended to be a separation by molecular size in solution followed by a separation according to composition with mixed chromatographic mechanisms acting synergistically. The complications encountered are discussed and the status of OC in light of recently published results is examined.

Polymers are generally highly complex, multicomponent materials. The molecules present can vary in molecular weight, composition, sequence length (the number of one type of monomer unit in a row before another monomer type is encountered), branch length, branch frequency, and tacticity. The concentration of each type of molecule present often dictates product performance. Frequently molecular weight is the property of interest and size exclusion chromatography (SEC) is used to analyze the polymer. SEC analysis is directed at obtaining a fractionation with respect to molecular weight and at detecting the concentration of each different molecular weight present. The molecular weight distribution and molecular weight averages can then be calculated from the chromatogram. This approach is quite straightforward for linear homopolymers where the only diversity in molecules originates from the differences in molecular weight (Figure 1). However, if the polymer is a linear copolymer, for example, instead of a linear homopolymer, then molecules can vary

0097–6156/87/0352–0059$06.00/0
© 1987 American Chemical Society

in composition and sequence length as well as in molecular weight
(Figure 2). It is very likely that these two additional property
distributions also significantly affect product performance.
Therefore, they too should be elucidated. Furthermore, even if it is
decided to ignore their possible effect on performance and instead
continue to focus on molecular weight, these other distributions can
interfere with the molecular weight analysis. This situation leads
us to the subject of analysis of "complex polymers" and the
development of a method to accomplish it, orthogonal chromatography
(OC).

In the next section, the term "complex polymers" is defined, the
effects of polymer complexity on conventional SEC analysis are
examined, and attempts to analyze complex polymers by utilizing SEC
detector technology are summarized. High performance liquid
chromatography (HPLC) attempts to accomplish the task are then
described. This is followed by a summary of the theoretical
development of OC, experimental results of OC analysis, complications
which emerged, and finally a summary of the status of OC in light of
recent developments. (1) and (2) provide reviews of OC.

SEC Analysis of Complex Polymers

Definition of a Complex Polymer. A simple polymer is one which has
at most one broad molecular property distribution (e.g., a broad
molecular weight distribution). A complex polymer is one which has
two or more broad molecular property distributions (e.g., a broad
molecular weight distribution and a broad copolymer composition
distribution)(3). Properties such as molecular weight and
composition, which can be in so much variety in a polymer that they
must be described as a distribution, are here termed "distributed
properties". It is the presence of simultaneous breadth (i.e.,
variety) in more than one distributed property which is the defining
characteristic of a "complex" polymer and the source of analysis
difficulties.

In this paper, typical linear homopolymers and linear copolymers
(as shown in Figures 1 and 2, respectively) exemplify simple and
complex polymers.

Effects of Polymer Complexity on Conventional SEC Analysis. In SEC
analysis of complex polymers both fractionation and detection can be
adversely affected. With regards to fractionation, the difficulty
originates from the fact that SEC separates on the basis of molecular
size in solution. This is synonymous with a molecular weight
separation only if there is a unique relationship between molecular
size and molecular weight. Although this is the case for linear
homopolymers it is generally not so for complex polymers such as
linear copolymers. For example, various combinations of molecular
weight, composition and sequence length can combine to yield the same
molecular size in solution. Figure 3 illustrates this by showing the
result of an ideal SEC separation for three different samples: a
linear homopolymer; a blend of two linear homopolymers; and a linear
copolymer. Thus, the presence of property variations in properties
other than molecular weight confounds interpretation of the
fractionation.

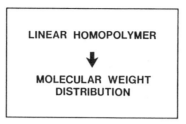

$$
\begin{bmatrix} -CH-CH_2- \\ \bigcirc \end{bmatrix}_n
$$

A

A~AAAA~A
A ~AA~A
A~A~A

Figure 1: Property distribution in a linear homopolymer: molecular weight distribution of polystyrene (styrene units represented by "A").

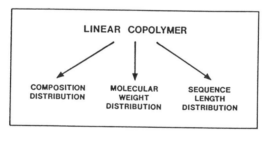

$$
\begin{bmatrix} -CH-CH_2- \\ \bigcirc \end{bmatrix}_n \quad \begin{bmatrix} CH_3 \\ -CH_2-C \\ O=C-O-CH_2-CH_2-CH_2-CH_3 \end{bmatrix}_m
$$

A   B

A~AABBBB~B
A~BABA~B
A~ABBA~A
A~AAA~A

Figure 2: Property distributions in a linear copolymer: composition distribution, molecular weight distribution and sequence length distribution of poly(styrene-co-n-butyl methacrylate). (Styrene units are represented by "A" and n-butyl methacrylate units by "B".)

It is worth noting at this point that Hamielec et al. (4) have
defined a complex polymer as any polymer not having a unique
relationship between molecular weight and molecular size. This is a
very useful definition although it can sometimes be too restrictive
(e.g., in describing a "complex polymer" for detection or for nonsize
exclusion mechanisms).

Complex polymers can also provide an ambiguous detector
response. For a linear homopolymer, if only one conventional
detector is present (e.g., a differential refractometer or a UV
detector set at one fixed wavelength) polymer concentration at any
retention time is determined from the proportionality of the detector
response to concentration. The detector is assumed to not respond to
differences in molecular weight (a good assumption for high molecular
weights). However, if a complex polymer such as a typical linear
copolymer is being analyzed, it must also be assumed that the
detector response is independent of composition and sequence length
as well while remaining proportional to concentration. This
assumption is generally invalid and detector response therefore
ambiguous to interpret. The single response can be attributed
simultaneously to polymer concentration, composition and sequence
length.

Attempts to Analyze Complex Polymers Using SEC Detector Technology.
For linear copolymers, multiple detectors and, more recently, diode
array UV/vis spectrophotometers have been used in attempts to
overcome the above analysis problems. The basic idea is to provide
more than one detector response so that the polymer concentration and
the number of properties will together equal the number of detector
responses (Figure 4). This provides the same number of equations as
the number of unknowns (5,6).

There are several difficulties associated with this approach.
The most fundamental one is that the fractionation problem associated
with polymer complexity is ignored. Therefore, for a complex
polymer, since each detector is really seeing a wide variety of
molecules, only average property values at each retention time can be
obtained and a polymer may appear to be uniform in composition when
there is actually significant composition variety.

In recent years other detectors have appeared. These include
low -angle laser light scattering and intrinsic viscosity. Again,
the fractionation problem is not dealt with and the same situation
holds as described above.

Attempts to Analyze Complex Polymers by High Performance Liquid
Chromatography (HPLC)

There have been many studies directed at using adsorption and
reversed-phase HPLC to separate copolymers by composition (1-3). Two
interacting problems associated with these approaches are:
    o The presence of one property distribution interferes with
separation on the basis of the other. For example, in adsorption
chromatography, the degree of adsorption can be affected by both the
molecular weight and by the composition of the molecule. For a
linear copolymer, adequate fractionation requires that the
composition differences completely dominate.

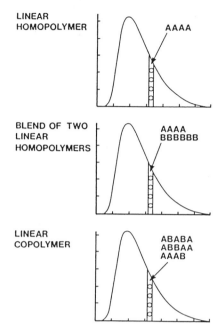

Figure 3: SEC Fractionation (letters A and B refer to different types of monomer units) (Reproduced with permission from Ref. 3. Copyright 1987, John Wiley & Sons.)

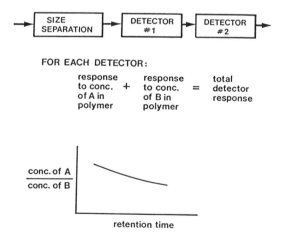

Figure 4: Dual detector SEC: Solution of the two equations for the two unknown concentrations leads to plots of the concentration ratio versus retention time. (Reproduced from Ref. 6. Copyright 1983, American Chemical Society.)

o Mixed separation mechanisms can cause the result to be
unpredictable. For example, large pores that allow polymer molecules
to enter so that large packing areas are available for adsorption can
cause a size exclusion separation to be superimposed upon an
adsorption separation. Small pores that exclude the molecules
severely limit the available packing area. Adequate fractionation
requires that one mechanism dominates the situation or that different
mechanisms can somehow be forced to act in the same direction.

Some studies have successfully demonstrated separations of
linear copolymers using HPLC. However, it is important to realize
that unlike in the HPLC of small molecules where a peak shows the
concentration of only one type of molecule, the SEC chromatogram of a
complex polymer is really an envelope covering possibly thousands of
different components. Even with modern detectors it is often very
difficult to ascertain that the desired fractionation has really been
accomplished. Universal calibration in SEC is of practical utility
because the same fractionation (i.e., a fractionation according to
molecular size) occurs whether monodisperse, polydisperse or complex
polymer molecules are involved. The fractionation is reliable. This
is a very difficult requirement for HPLC methods because of the
variety of complex molecules that can be present.

Theoretical Development of OC

Orthogonal chromatography (OC) is a method of using SEC to analyze
complex molecules (2,6). To accomplish the desired fractionation,
two SEC instruments are coupled together so that the eluent from the
first flows through the injection valve of the second. Figure 5
shows this arrangement. The first instrument is operated as a
conventional SEC. The second is operated with a solvent/nonsolvent
mixture to utilize nonsize exclusion mechanisms. The desired
detection is attained by utilizing new detector technology, notably a
diode-array UV/vis spectrophotometer, at the exit of the second
instrument. The basic principles underlying OC are discussed in the
following paragraphs.

Cross-Fractionation. Complex polymers contain more than one broad
property distribution. If molecular weight and composition are the
only two property distributions present then an example of
cross-fractionation would be the separation of the polymer first
according to molecular weight and the separation of each single
molecular weight fraction obtained according to composition. This
cross-fractionation provides a two-dimensional answer to a
two-dimensional distribution problem. It has typically been
accomplished for polymers using solvent/non-solvent precipitation.

Multi-dimensional Chromatography. Multi-dimensional chromatography
is the term used to describe a variety of methods where fractions
from one chromatographic system are each transferred to another for
further separation. Combinations of SEC with thin-layer
chromatography have been shown to enable separation of copolymers by
composition in a "cross-fractionation". OC utilizes a combination of
two SECs in a cross-fractionation approach.

Synergistic Use of Mixed Separation Mechanisms. In OC the first SEC
utilizes conventional separation by molecular size exclusion.
However, the presence of the nonsolvent in the mobile phase of the
second instrument encourages adsorption/partition effects as well as
size exclusion there.

Figures 6 and 7 illustrate the proposed mechanism in OC. Using
the specific example of a separation of a styrene n-butyl
methacrylate copolymer, the first SEC separates the copolymer
according to molecular size in solution. At any desired retention
time, the flow in the first instrument is stopped and an injection
made into the second instrument of a single molecular size "slice" of
the chromatogram. The solvent running in the second instrument is a
mixture of tetrahydrofuran (THF) and n-heptane. THF is a solvent for
both styrene and n-butyl methacrylate portions of the polymer
molecules. However, n-heptane is a nonsolvent for the styrene-rich
portions. As a result, when the injection is made into the second
instrument, the styrene-rich molecules will shrink relative to the
n-butyl methacrylate-rich molecules. An immediate size distribution
will be present which will reflect the composition differences. The
smaller styrene-rich molecules will enter more pores of the column
packing than their n-butyl methacrylate-rich counterparts and so be
fractionated. Furthermore, since the styrene-rich molecules "hate"
the mobile phase, they should find the surface area of the packing
more "sticky" than the n-butyl methacrylate-rich molecules. Thus,
again the styrene-rich molecules should be retarded relative to the
others. According to this picture, the mechanisms of size exclusion,
adsorption and partition are thus able to act synergistically to
accomplish a composition separation.

Detector Technology. For copolymer composition analysis the new
diode array UV/vis detectors are extremely attractive: the
absorption at many wavelengths are instantaneously recorded; there
is only a single spectrophotometer cell so that transport time delays
between detectors and axial mixing in detector cells do not confound
comparison of detector response at different wavelengths; and for
styrene copolymers, extremely low concentrations of polymer can be
detected.

Results

Initial Work. OC was developed by injecting solutions containing
mixtures of polymers into the first SEC (5). Polystyrene,
poly(n-butyl methacrylate) and poly(styrene co-n-butyl methacrylate)
were used. Initially the main objective was to demonstrate that a
composition separation was actually being obtained and that the
results were not simply some artifact of sampling mixtures of
polymers with different molecular weight distributions. This was
shown by running a variety of different samples and by actually
analyzing the spread of molecular weights in a chromatogram "slice"
by running the second instrument with pure THF instead of the usual
THF/ n-heptane mixture. At this point in the development twelve SEC
columns were used in the first SEC and three in the second. One hour
was required to obtain the first analysis. Separation of polystyrene
from poly(styrene co-n-butyl methacrylate) in a blend provided the
best results (5, 7).

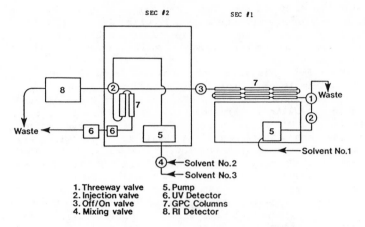

1. Threeway valve   5. Pump
2. Injection valve   6. UV Detector
3. Off/On valve   7. GPC Columns
4. Mixing valve   8. RI Detector

Figure 5: Arrangement of SEC instruments in Orthogonal
Chromatography. Note that the 9 to 12 columns used for SEC 1 were
reduced to 3 in later work to reduce analysis times. (Reproduced
from Ref. 5. Copyright 1980, American Chemical Society.)

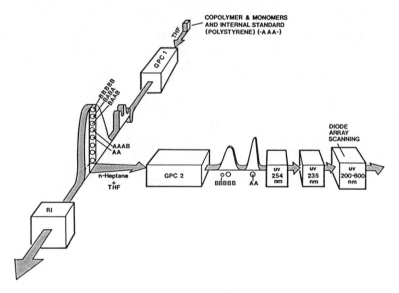

Figure 6: Schematic diagram of Orthogonal Chromatography showing
size fractionation of a linear copolymer by SEC 1, the variety of
molecules of the same molecular size within a chromatogram "slice"
(in this case A refers to styrene units and B to n-butyl
methacrylate units) and composition fractionation by SEC 2.
(Reproduced from Ref. 6. Copyright 1983, American
Chemical Society.)

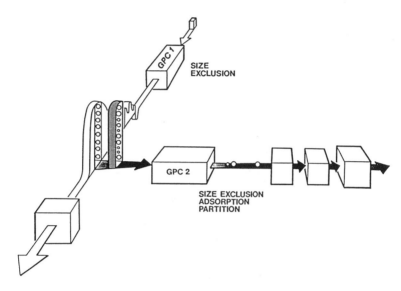

Figure 7:   Schematic diagram of Orthogonal Chromatography showing
the separation mechanisms involved.   (Reproduced
from Ref. 6.   Copyright 1983, American Chemical Society.)

Axial Dispersion Characterization. Use of THF in both instruments as a method of examining the fractionation situation led to the investigation of OC as a method of supplying polymer of extremely narrow molecular weight distribution for resolution characterization of the second instrument (7). To do this, a commercially available narrow molecular weight distribution standard was injected into the first instrument and sampled at its peak by the second instrument.

Poly(Styrene co-n-Butyl Methacrylate) Fractionation. OC was developed with the particular idea of elucidating the kinetics of the free radical copolymerization of styrene n-butyl methacrylate. Thus, this polymer provided the main focus of the work.

Figure 8 (1, 6) shows the fractionation obtained by analyzing a mixture of polystyrene, poly(styrene co-n-butyl methacrylate) and poly(n-butyl methacrylate) with various n-heptane concentrations. Once it was realized that multiple columns in the first SEC really did not offer any advantage in terms of greater injection amounts because of increased dilution in the columns, smaller injections and less columns reduced analysis times by 50% with no loss in sensitivity. For the analyses shown in Figure 8, only three columns were used in the first SEC and three in the second. With this system the first analysis by both SEC instruments required a total of 30 minutes and subsequent analyses of the same sample about 15 minutes each. However, despite these significant reductions in analysis times in comparison to the initial work, complete analysis of even one complex polymer required many cross fractionations and generated much data.

The diode array UV/vis spectrophotometer was used to both identify the polymer exiting and to obtain a quantitative analysis of the copolymer composition distribution. Figure 9 (6) shows the result of summing many individual fraction analyses to see the total copolymer composition distribution. The result had the correct average composition but not the skewed shape expected from theory. Part of the difficulty was the relatively small number of cross fractionations done.

OC of Other Polymers. Figure 10 shows the fractionation of poly(ethyl methacrylate), polystyrene and poly(lauryl methacrylate). The THF/n-heptane mobile phase worked well for these polymers and UV absorptivity was sufficient for detection. The detection requirement was especially important because of the low concentrations resulting in OC.

Complications

o SEC calibration curves obtained in the usual way by injecting "monodisperse" polystyrene standards directly into the second SEC using various mobile phases in that SEC showed a peculiar "jump" in retention time when n-heptane concentration was changed from 57 to 60%. These calibration curves also demonstrated that higher molecular weight polystyrenes were greatly retarded by the n-heptane concentration (Figure 11).

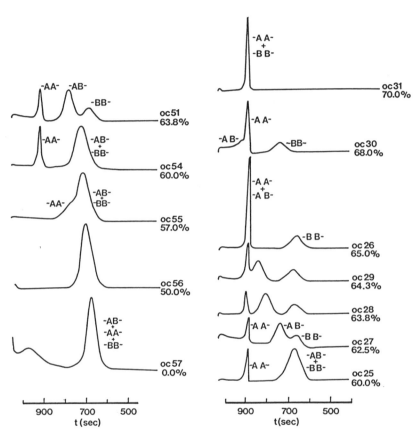

Figure 8: Two series of Orthogonal Chromatography runs showing the effect of % n-heptane in the mobile phase of SEC 2 on the fractionation (AA: polystyrene, AB: poly(styrene co-n-butyl methacrylate); BB: poly n-butyl methacrylate). (Reproduced from Ref. 6. Copyright 1983, American Chemical Society.)

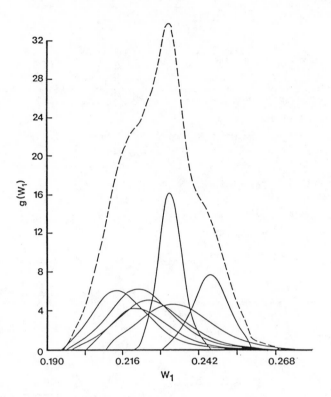

Figure 9:   Copolymer composition distribution for whole polymer as sum of distributions obtained from individual cross fractionations.   (Reproduced from Ref. 6.   Copyright 1983, American Chemical Society.)

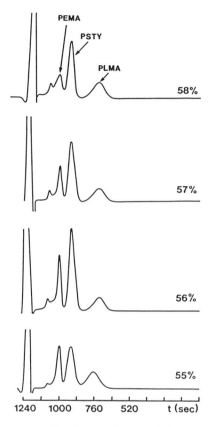

Figure 10: Separation of a homopolymer mixture of poly (ethyl methacrylate) (PEMA), polystyrene (PSTY) and poly (lauryl methacrylate) (PLMA) by Orthogonal Chromatography at different % n-heptane concentrations in SEC 2.

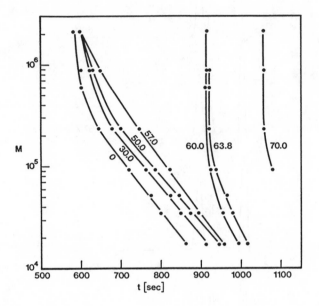

Figure 11: Polystyrene calibration curves for SEC 2: variation
with % n-heptane. (Reproduced from Ref. 6. Copyright 1983,
American Chemical Society.)

o Cloud point experiments showed that some of the higher molecular weight polystyrene standards used in analyses should have been precipitating at n-heptane concentrations beyond 65%.
o The resolution of the second SEC was strongly affected by the order of columns. Figure 12 shows the result of changing the position of the smallest pore size column from the furthest from the injection valve, to closest, when analyzing a blend of polystyrene and poly(n-butyl methacrylate).

Status of OC

Effect of Injected Solvent. It was eventually determined that the mobile phase injected from the first SEC (i.e., pure THF) affected the separation in the second SEC. This is dramatically demonstrated in Figure 13 which shows the result of injecting a narrow molecular weight distribution polystyrene sample directly into the second SEC. Mobile phase was of constant composition throughout (63.8% n-heptane in THF). However, the solvent used to dissolve the polystyrene was varied from 0% n-heptane in THF to 50% and plotted on the abscissa versus peak retention time on the ordinate. Peak retention time varied from 915 seconds at 0% n-heptane to 960 seconds at 50%.
The consequences of this effect of injected solvent are as follows (6):
o   Operation of the second SEC is effectively always in a gradient mode. The "plug" of THF injected with the polymer sweeps through the columns to form this gradient. At higher n-heptane content mobile phases, styrene-rich polymers would prefer to remain with this THF plug. However, small pore size packing causes the THF to be separated from the styrene-rich polymer because the THF enters more pores. If this packing is remote from the detector the THF may capture this polymer later in larger size packings since the polymer is also retarded by nonexclusion mechanisms (e.g., adsorption). If the packing is near the detector then the styrene-rich polymer may be separated from the THF pulse long enough to exit separately.
o   Precipitation of styrene-rich polymer is a possibility. No column plugging or significant loss of polymer to the columns (except for a few conditions) was observed. This could be due to the sweeping of the columns by the THF pulse. However, since no precipitation was evident in the detector and since the styrene-rich polymers were able to be separated from the THF pulse (see previous paragraph) it is likely that solvation of the polymer by the THF is affecting the results.
o   Sequence length can be affecting both fractionation and detection. Fractionation in the first SEC (according to molecular size) is "universal". However, the composition fractionation in the second SEC may be scrambled by sequence length variations. There is some evidence that sequence length affects UV detector response. Thus, a diode array spectrophotometer could be used to obtain sufficient information to elucidate both composition and sequence length.

Recent Advances Relevant to OC. HPLC studies of polymers increasingly employ columns with very small pores to purposefully exclude all polymer molecules (8-13). The reasons justifying

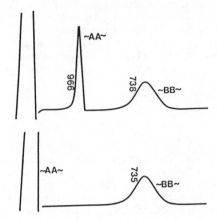

Figure 12: Effect of column reordering on polystyrene (AA) and
poly (n-butyl methacrylate) (BB) retention in SEC 2. Key: top,
small pore size column last; and bottom, small pore size column
first. (Reproduced from Ref. 6. Copyright 1983,
American Chemical Society.)

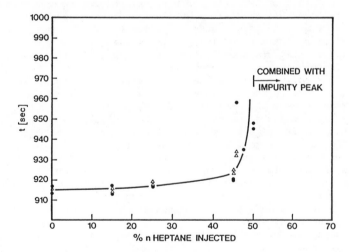

Figure 13: Effect of injected solvent composition on polystyrene
retention in SEC 2. (Reproduced from Ref. 6. Copyright 1983,
American Chemical Society.)

this approach are: The high external surface areas of high resolution packings (estimated at 20% of the total surface area of a packing material with pores large enough to accommodate all of the polymer molecules) (10); the possible ability of polymer molecules to uncoil sufficiently to be affected by small pores in adsorption-type mechanisms (12).

Although there remains some disagreement (12,13), the presence of a precipitation mechanism in separations using gradient chromatography of polymers in HPLC separations appears very likely (8,9). This reinforces the hypothesis mentioned above that precipitation is also an important mechanism in separation in the second SEC used in OC. However, it is important to note that considering the complexity of the polymers to be analyzed and the variety of important packing variables, mixed mechanisms must be anticipated. Attempting to arrange fractionation to synergistically use these mechanisms appears as a more reliable and more generally applicable approach then depending upon the dominance of any one mechanism. However, it is now evident that even then, the potential polymer complexity and current uncertainty associated with these mechanisms demands use of modern detector technology to identify the exiting molecules as thoroughly as is required by the purpose of the analysis.

Glockner et al. (9) have shown very good separations of styrene acrylonitrile copolymers using an SEC and a HPLC set up as are the two SEC's in OC. They term the separation occurring in the HPLC "high performance precipitation liquid chromatography".

The retarding effect of pores on solvent relative to the polymer described in Section 6.1 has now been proposed as the reason for the good separations obtained by precipitation mechanisms (8-10, 13). The polymer is visualized to continually re-precipitate and re-dissolve as the solvent front of a gradient repetitively overtakes and then loses the polymer.

Garcia Rubio et al. (14,15) have accomplished significant development in the UV analysis of copolymers. In examining the data justifying the use of UV spectra for determining both composition and sequence length of polymers, application of error propagation theory showed that the published results on measurement of sequence length by UV can be explained by considering only composition, at least for certain wavelengths. Furthermore, they showed that UV absorption spectra are significantly affected by benzoate groups on the polymer. These groups are produced during polymerization when benzoyl peroxide is used as the initiator. For accurate quantitative work the spectra must be corrected for this absorption. Some of the inaccuracy in Figure 9 could well be due to this source of error.

Much UV development work utilizes "off-line" analysis (detector not connected to an SEC) of precipitated polymer. It involves very detailed interpretation of differences in spectra. One caution which must be observed is to ensure that residual monomer or other small molecules are not causing spurious results. Acquiring spectra "on-line" using a diode array UV/vis spectrophotometer attached to an SEC is one answer to this problem. However, then determining the concentration of polymer examined can be troublesome.

Furthermore, it should be noted that interpreting polymer spectra in solvent mixtures typical of gradient operation in liquid chromatography may be an additional problem. Although all solvents

may be transparent to UV, the conformation of the polymer is expected
to change with solvent composition and this in turn can affect the
observed spectra.

Conclusions

o Conventional SEC analysis of complex polymers to elucidate
individual property distributions (other than the distribution of
molecular sizes) encounter difficulties in both effective
fractionation and unambiguous detection.

o The use of new detector technology has been the main emphasis in
attempts to use SEC for the analysis of complex polymers.

o HPLC attempts to analyze complex polymers have emphasized
fractionation employing gradient operation and adsorption or reversed
phase packings. Recent notable work includes the use of SEC with
HPLC.

o OC is a multi-dimensional SEC method which emphasizes both
fractionation and detection.

o OC has been successfully applied to accomplish a composition
separation of styrene/methacrylate homopolymers and copolymers.

o A mechanism has been postulated to account for the observed
separations and proposes that the various mechanisms involved can act
synergistically.

o The participation of the mobile phase from the first SEC in the
separation observed in the second SEC effectively creates a gradient
operation in the second.

o Successive precipitation and dissolution of the polymer in the
second SEC is likely an additional important separation mechanism.

o Quantitative analysis has been attempted but detection of sequence
length is needed.

o Development of UV detector interpretation and increased automation
of column switching and data acquisition/interpretation are important
to future OC development.

Literature Cited

1. Balke, S.T. "Quantitative Column Liquid Chromatography, A
    Survey of Chemometric Methods"; Elsevier: Amsterdam, 1984.
2. Balke, S.T. Sep. Purif. Methods 1982, 11, 1.
3. Balke, S.T. In "Modern Methods of Polymer Analysis"; Barth,
    H.G., Ed.; Wiley: New York, 1987 (in press).
4. Hamielec, A.E.; Styring, M. Pure & Appl. Chem. 1985, 57, 955.
5. Balke, S.T.; Patel, R.D. In "Size Exclusion Chromatography
    (GPC)"; Provder, T., Ed.; ACS SYMPOSIUM SERIES No. 138, American
    Chemical Society: Washington, DC, 1980; p. 149.

6. Balke, S.T.; Patel, R.D. In "Polymer Characterization:
    Spectroscopic, Chromatographic, and Physical Instrumental
    Methods"; Craver, C.D., Ed.; ADVANCES IN CHEMISTRY SERIES No.
    203, American Chemical Society: Washington, DC, 1983; p. 281.
7. Balke, S.T.; Patel, R.D. <u>J. Polym. Sci., Polym. Letters</u> 1980,
    18, 453.
8. Glockner, G. <u>TRAC</u> 1985, 4, 214.
9. Glockner, G.; Van Den Berg, J.H.M.; Meijerink, N.L.J.; Scholte,
    T.G.; Koningsveld, R. <u>J. Chromatogr.</u>, 1984, 317, 615.
10. Glockner, G. <u>Pure & Appl. Chem.</u> 1983, 55, 1553.
11. Mourey, T.H.; Noh, I.; Yu, H. <u>J. Chromatogr.</u> 1984, 303, 361.
12. Snyder, L.R.; Stadalius, M.A.; Quarry, M.A. <u>Anal. Chem.</u> 1983,
    55, 1412A.
13. Armstrong, D.W.; Boehm, R.E. <u>J. Chromatogr. Sci.</u> 1984, 22,
    378.
14. Garcia-Rubio, L.H.; Ro, N.; Patel, R.D. <u>Macromolecules</u> 1984,
    17, 1998.
15. Garcia-Rubio, L.H.; Ro, N. <u>Can. J. Chem.</u> 1985, 63, 253.

RECEIVED March 10, 1987

DETECTION

Chapter 5

A New Stand-Alone Capillary Viscometer Used as a Continuous Size Exclusion Chromatographic Detector

W. W. Yau, S. D. Abbott, G. A. Smith, and M. Y. Keating

Central Research and Development Department,
E. I. du Pont de Nemours & Co., Wilmington, DE 19898

This paper describes a new design of a forced–flow–
–through–type capillary viscometer used for batch
sample viscosity measurements as well as continuous
viscosity detection for size exclusion chromatography
(SEC). In one version of our viscometer design, an
analytical capillary is connected in series with a
reference capillary in the flow stream and the
pressure drop across the capillaries is measured by
pressure transducers. A differential log–amplifier is
used to convert the two transducer signals into an
output signal that is directly proportional to the
natural logarithm of the relative viscosity of the
sample fluid. The output signal is highly insensitive
to flow rate fluctuations and thus gives a very
sensitive and accurate means to measure viscosity.
The sample fluid could be any neat liquid or a sample
of polymer solution. Under favorable conditions, a
single viscosity determination on a polymer solution
at high dilution can provide a direct measure of the
polymer intrinsic viscosity, without the need of
polymer concentration extrapolation. With this
viscometer used as a continuous viscosity detector for
SEC, it is possible to achieve SEC molecluar weight
calibration by way of the universal SEC calibration
methodology without the need of molecular weight
standards for the unknown polymers.

Background

Accurate measurements of fluid viscosity are important in many
industries for such diverse uses as monitoring syrup manufacture or
studying polymer structures such as polymer branching, chain
conformation, solvent interactions or polymer molecular weight (MW).
Historically, the drop–time type glass capillaries, such as the
Ubbelohde or Cannon and Fenske types, have been widely used to
measure fluid viscosity. However, this traditional method is tedius
and labor intensive, and lacks the desired speed and sensitivity to

0097–6156/87/0352–0080$06.75/0
© 1987 American Chemical Society

meet the needs of viscosity measurements, especially for dilute polymer solution characterizations.
The following is a brief review of the viscosity parameters that are commonly used in polymer analyses. The relative viscosity (η_{rel}) of a polymer sample solution as defined in Equation 1 can be determined experimentally from the measured viscosity value for the polymer sample solution (η) and that of the solvent (η_o). From the η_{rel} value and the polymer sample concentration (c), the calculations for the other viscosity parameters are possible in accordance to Equations 2 through 5:

Relative Viscosity: $\qquad \eta_{rel} = \eta/\eta_o \qquad$ (1)

Specific Viscosity: $\qquad \eta_{sp} = \eta_{rel} - 1 \qquad$ (2)

Inherent Viscosity: $\qquad \eta_{inh} = (\ln \eta_{rel})/c \qquad$ (3)

Reduced Viscosity: $\qquad \eta_{red} = \eta_{sp}/c \qquad$ (4)

Intrinsic Viscosity: $\qquad [\eta] = \lim_{c \to 0} \eta_{inh} = \lim_{c \to 0} \eta_{red}$ (5)

where the mathematical symbol ln means natural logarithm, and $\lim_{c \to 0}$ means the limiting value for the viscosity parameter as the sample concentration c approaches zero at infinite dilution.
The experimental determination of polymer intrinsic viscosity is done through the measurement of polymer solution viscosity. The connotation of intrinsic viscosity [η], however, is very different from the usual sense of fluid viscosity. Intrinsic viscosity, or sometimes called the limiting viscosity number, carries a far more reaching significance of providing the size and MW information of the polymer molecule. Unlike the fluid viscosity, which is commonly reported in the poise or centipoise units, the [η] value is reported in the dimension of inverse concentration units of dl/g, for example. The value of [η] for a linear polymer in a specific solvent is related to the polymer molecular weight (M) through the Mark–Houwink equation:

$$[\eta] = KM^\alpha \qquad (6)$$

where K and α are Mark–Houwink viscosity constants, some of which are available in polymer handbooks.[2] The usual value for α falls between 0.5 and 0.8 for polymers of the random–coil type conformation in solution.
A more sensitive viscometer than the drop–time glass capillary method is also needed in size exclusion chromatography (SEC) such as the gel permeation chromatographic (GPC) analysis of polymer molecular weight distribution (MWD). In an SEC system, a concentration detector is commonly used for providing the weight concentration profile of the polymer elution curve. The MW and MWD information of the sample is provided indirectly by the retention time of the different polymer components in the sample. Quite obviously, it is highly desirable to have additional detectors available for SEC that are sensitive to the molecular weight of the different polymer components eluting from

the SEC columns. The on-line viscosity and light scattering detection[3] capabilities in addition to the usual SEC concentration detector are very useful in achieving absolute SEC-MW calibration. The required features of such a MW-specific detector for SEC include: continuous mode of monitoring, high sensitivity, low noise, and low dead volume for minimal SEC band broadening.

Earlier experiments[4] involved the collection of SEC effluent aliquots to measure solution viscosity in batches with the very time consuming Ubbelohde drop-time type viscometers. A continuous capillary type viscometer was first proposed for SEC by Ouano[5]. Basically, as shown in Figure 1, a single capillary tube with a differential pressure transducer was used to monitor the viscosity of SEC effluent at the exit of the SEC column. As liquid continuously flows through the capillary (but not through the pressure transducer), the detected pressure drop (ΔP) across the capillary provides the measure for the fluid viscosity (η) according to the Poiseuille's viscosity law:

$$\Delta P = k\ Q\ \eta \qquad (7)$$

where Q = flow rate, k = capillary geometrical constant,

$$k = 8L/\pi R^4 \qquad (8)$$

with L = capillary length, and R = capillary inside radius. Initial testing of such a detector was encouraging mainly because of its capability to continuously detect and record the SEC-viscosity elution profile. The detector fell short of being entirely successful due to unfavorable signal-to-noise problems. There have been other attempts at the SEC viscosity detector based on single capillary design.[6,7] The performance of these viscometers, however, remains marginal because the pressure drop ΔP signal of a single capillary is highly subjective to the unavoidable flow rate and temperature fluctuations.

Attempts to compensate for flow rate fluctuation have been made by the use of multiple capillary tubes.[8-10] Both Blair's and Haney's viscometers use a parallel bridge design of four capillaries.[8,9] The desired fluid viscosity response is detected by measuring the differential pressure across the capillary bridge. Haney's design has led to a commercial capillary viscometer (Viscotek Corp., Porter, Texas). The device provides much superior sensitivity over the earlier single capillary designs. Flow rate and temperature fluctuations are largely eliminated to provide a very stable baseline. The size of the sample viscosity signal, as measured by the differential pressure across the capillary bridge, however, is still highly affected by flow rate changes. The measured differential pressure is directly proportional to the overall flow rate across the capillary bridge. The series capillary design of Abbott and Yau overcomes the problem of flow rate dependency of the viscometer response.[10] This latter viscometer design is the subject of this paper. While this viscometer is described herein, with reference particularly to polymer-solvent solutions, it should be noted that the viscometer may be used with other type sample liquids as well.

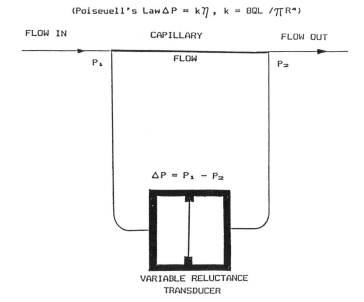

Figure 1. Differential Pressure Detection of Fluid Viscosity.

The Key Design Features

The new viscometer design utilizes two sets of the capillary and pressure transducer assemblies like the one shown in Figure 1. The capillaries are connected in series as shown in Figure 2 for the stand-alone viscometer configuration. At the time the sample solution passes through the analytical capillary, the reference liquid or the carrier solvent continuously flows through the reference capillary. The differential pressure signals from the two capillary-transducer systems are fed to a differential logarithmic amplifier. The differential logarithmic amplifier compares the input signals and gives the ratio between the analytical and the reference pressure drops $\Delta P_A / \Delta P_R$. The output of the log-amplifier gives the desired measure of the natural logarithm of the sample relative viscosity, that is $\ln \eta_{rel}$. Real time signal processing of the simultaneous pressure drops across the analytical and reference capillaries eliminate the effects of flow rate and temperature fluctuations from the log-amplifier output signal. The elimination of the flow rate noise is the key to the high sensitivity and the signal-to-noise performance of the present viscometer. The following is a mathematical analysis of the differential log-amplifier output signal (s) of the viscometer with the series capillary design:

$$S = \ln \Delta P_A - \ln \Delta P_R$$

$$= \ln (\Delta P_A / \Delta P_R)$$

$$= \ln (G_A k_A Q_A \eta / G_R k_R Q_R \eta_0) \qquad (9)$$

where G is the electronic gain, k is the capillary geometrical constant as defined before in Equation 8, Q is again the flow rate, η is the viscosity of the fluid in the analytical capillary, and η_0 is the solvent viscosity; the subscripts A and R refer to the analytical and the reference capillary, respectively. Under the usual solvent flow rate and viscosity conditions of SEC and viscometric measurements, the laminar flow requirement of the Poiseuille's equation (Eq. 8) is easily satisfied. The solution flow through the measuring capillary usually does not exceed a Reynold's number of 100, far below the condition for the on-set of turbulence. Since the flow rates in two capillaries connected in series have to be the same, that is $Q_A = Q_R$, flow rate effects are cancelled out from Equation 9 to give the flow rate independent signal:

$$S = \ln (G_A k_A \eta / G_R k_R \eta_0)$$

$$= \ln \eta_{rel} + \ln (G_A k_A / G_R k_R) \qquad (10)$$

The second term in Equation 10 is a zero offset factor for the desired viscometer signal of $\ln \eta_{rel}$ of the first term, where $\eta_{rel} = \eta / \eta_0$ as noted before.

STAND-ALONE VISCOMETER

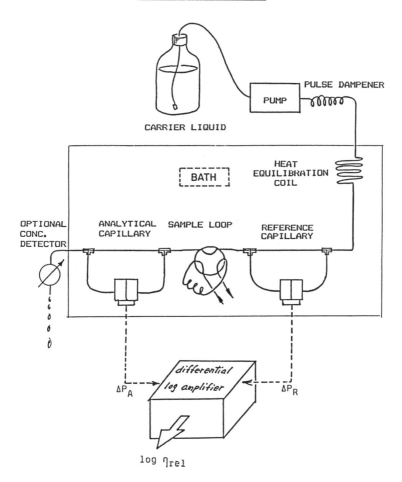

Figure 2. Viscometer-Configuration 1.

By adjusting the gains of the ΔP signal electronically so as to match $G_A k_A$ and $G_R k_R$, that is $\ln(G_A k_A/G_R k_R) = \ln 1 = 0$ to give the direct $\ln \eta_{rel}$ readout:

$$S = \ln \eta_{rel}$$
$$= \eta_{inh} \times c \qquad (11)$$

And, at sufficient sample dilution,

$$S = [\eta] \times c \qquad (12)$$

The electronic matching of the capillary performances can be easily accomplished by nulling the log-amplifier output when the viscometer pumps solvent through both the analytical and the reference capillaries, that is when $\eta = \eta_0$ and $\ln = \ln 1 = 0$.

The unique advantages of this new viscometer design include: (1) true $\ln \eta_{rel}$ readout independent of flow rate, (2) no need to match the capillary flow resistances and the transducer response factors between the analytical and the reference capillary, (3) quick and convenient rematching of capillary performance is easily done electronically to offset any long term drift of the capillary resistance due to the polymer build-up on capillary walls or any response factor variations of the pressure transducers, (4) high accuracy over a wide dynamic range of 0.0001 to about 5 in the relative viscosity units. On the other hand, the viscometer is equally matched with the Viscotek viscometer in enjoying the following additional advantages: (1) high sensitivity of better than the 10^{-4} relative viscosity units, (2) due to the high sensitivity, single point $[\eta]$ determination is possible without the need of sample concentration extrapolation, (3) capillaries are of high length-to-diameter ratio and require no kinetic energy and end-effect corrections, (4) shear-rate can be controlled and varied to study the non-Newtonian behavior of a polymer solution or other neat liquid samples, (5) easy adaptation to automation, (6) option to add a concentration detector to the batch viscometer to allow in-situ sample concentration monitoring.

Configurations As a Stand-Alone Viscometer

Figures 2 and 5 illustrate two different configurations of the viscometer that can be used in batch sample viscosity determinations.

In the viscometer-configuration 1 shown in Figure 2, the reference capillary is placed before the sample injection valve. Carrier solvent is in the reference capillary all the time. Sample solution is introduced into the solvent flow stream from a sample loop via a sample injection valve. The solvent flow pushes the sample solution through the analytical capillary where the viscosity of the sample is detected. The ΔP signal from each pressure transducer is fed to a differential logarithmic amplifier as shown in the Figure. The viscometer output is the $\ln \eta_{rel}$ signal of the sample solution. A pump is used to circulate the solvent through the viscometer. The viscometer capillaries are immersed in a temperature controlled liquid bath

A concentration detector such as the differential refractometer is shown here connected in series with the capillaries. The following components are typically used in the viscometer: stainless steel capillaries of 1/16-in. o.d. and 0.016 in. i.d. X 8 in. long, 2 ml. sample loop, Celesco pressure transducers of 1 psi rating, Valco 6 port sample valve, Burr Brown Log 100 JP. type differential log-amplifier, VWR-1145 circulation temperature bath (-15 to 150°C). Several liquid chromatographic pumps have been used. A Du Pont 860 pump was used to obtain the data reported in this work.

In operation, the viscometer of Figure 2 will generate two separate signal detector traces for recording. The differential log-amplifier will generate a viscosity (ln η_{rel}) trace while the concentration detector will generate a concentration (c) trace. Both will occur simultaneously and repeatedly from successive sample injections as shown in Figure 3 for a polystyrene sample. From the ln η_{rel} and the c signals, the inherent viscosity of the polymer sample can be calculated directly and accurately from the ratio of the signal amplitudes shown in Figure 3. It is quite obvious that the use of in-situ sample concentration measurement has the advantage of reducing operator errors in preparing sample solutions of desired concentrations.

The flow rate independence of this viscometer has been demonstrated by intentionally varying the flow rate during the sample analyses. In Figure 4, the log-amplifier signal (ln η_{rel}) is recorded at the top, the ΔP_A and ΔP_R signals are also recorded as the middle and the bottom traces respectively. The upsets in the bottom ΔP_R trace reflect the intentional flow rate variations manipulated by upsetting the pump flow rate control. Such flow rate upsets have greatly disturbed the ΔP_A signal as well, especially at the top of the ΔP_A response to an injected sample viscosity. The log-amplifier signal, however, is not affected by the flow rate upsets. The integrity of the log-amplifier signal gives credence to the true ln η_{rel} measurement of the viscometer.

An alternative configuration for a batch viscometer is shown in Figure 5. In this case, the reference capillary is also placed downstream from the sample injection valve. A delay volume is added between the analytical and the reference capillaries. The function of the delay volume is to prevent sample solution reaching the reference capillary during the time that the sample viscosity is being monitored in the analytical capillary. With the delay volume, the sample ΔP_A signal is still referenced against the solvent ΔP_R signal to give the true ln η_{rel} measurement for the sample. At the completion of the sample measurement, the viscometer will reset itself as the sample solution flushes through the reference capillary. This viscometer-configuration 2 is shown in Figure 5 with an optional UV concentration detector connected in the parallel arrangement. In operation, the viscometer in Figure 5 will generate the dual trace viscosity and concentration signals shown in Figure 6 for a polyethylene-terephthalate sample. The flushing of the delay volume can be monitored by the returning of the negative log-amplifier signal back to baseline. Compared to the earlier viscometer configuration of no delay volume, this viscometer configuration offers a better signal-to-noise performance, however, at the cost of longer sample analysis time due to the additional time required to flush out the delay volume. A coiled large i.d. tubing of about 4 to 6 ml volume is typically used as the viscometer delay volume.

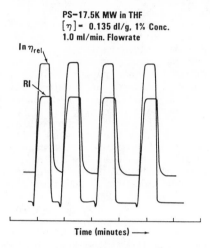

PRECISION OF STAND-ALONE VISCOMETER

PS-17.5K MW in THF
$[\eta] = 0.135$ dl/g, 1% Conc.
1.0 ml/min. Flowrate

Time (minutes) →

Figure 3. Precision of the Differential Pressure Viscometer.

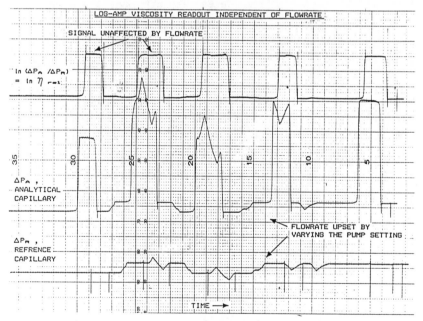

Figure 4. Flow rate Independent Viscometer Response.
Sample: polystyrene 17500 MW
Solvent: THF, concentration: 1%.

STAND-ALONE VISCOMETER

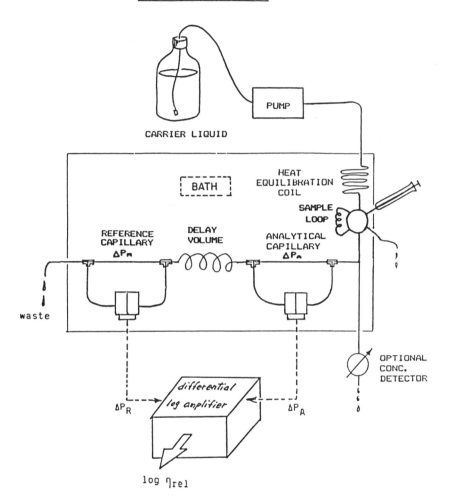

Figure 5. Viscometer-Configuration 2.

Due to the high sensitivity of the viscometer, accurate
readings of viscosity responses can be made with sample η_{rel} values
much less than 1.1. Therefore, sample solutions of very low
concentration can be analyzed in the viscometer to produce a single
point intrinsic viscosity determination without the need of
concentration extrapolation. As shown in Figure 7 for a
polystyrene sample, any η_{inh} determination at the η_{rel} value of
much less than 1.1 gives practically the intrinsic viscosity of the
sample. The approximation can be made even better by calculating
the sample intrinsic viscosity from the Solomon—Gatesman equation:

$$[\eta] = \lim_{c \to 0} \sqrt{2(\eta_{rel}-1-\ln\eta_{rel})}/c \tag{13}$$

Configurations as a SEC Detector

Figures 8 and 10 illustrate two different configurations of the
viscometer used as an on-line SEC viscosity detector.
    In the detector-configuration 1 shown in Figure 8, a SEC
column set is placed between the sample injection valve and the
analytical capillary of the viscometer. A large depository column
has been added between the analytical and the reference capillary.
The SEC concentration detector such as the differential refractometer (R.I.) is shown here connected in the series arrangement
following directly after the analytical capillary. Typically, a
large low pressure column of about 300 ml volume filled with large
packing beads can be used as the depository column.
    In operation, the SEC detector of Figure 8 will generate
both the viscosity (ln η_{rel}) and the concentration traces for
recording the SEC elution curve profile. The polymer bands eluting
from the SEC column and the analytical capillary will be drastically diluted in concentration when they finally emerge from the
depository column and reach the reference capillary. For all
practical purposes, the pressure drop across the reference
capillary will be responding to the solvent viscosity and any flow
rate changes, while the pressure drop of the analytical capillary
will respond to the viscosity of the eluting polymer bands as well
as any flow rate changes. Some build-up of polymer concentration
in the depository column will occur after many sample analyses in
close time intervals. Flushing out the depository column may be
necessary at times. This can be done for example by the continuous
pumping of solvent through the system overnight.
    The flow rate independence of the new SEC viscosity detector
design has been demonstrated by intentionally working with a very
large pump flow rate noise as shown in Figure 9. The noisy pump
response was created when two of the three pistons of a Du Pont 860
pump were made inoperative, leaving only a single reciprocating
piston to do the pumping. The Figure shows the ΔP_A and the ΔP_R
signals at two flow rate levels. while the SEC elution peaks are
barely visible in the noisy ΔP_A signal at the top trace, they are,
however, clearly detected in the log-amplifier signal shown at the
bottom of the Figure. This is the result of the very effective
cancellation of pump noise by the log-amplifier in the present
viscometer design. Another thing to notice is the size of the
elution peaks in the log-amplifier trace. The fact that the size

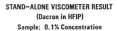

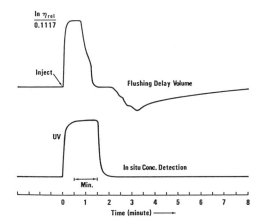

Figure 6. Typical Viscometer Output Signal Traces.

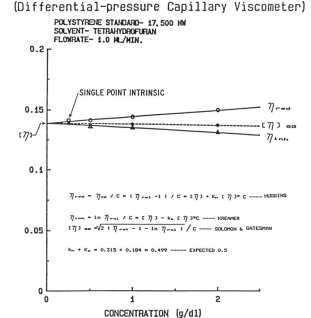

Figure 7. Single Point Intrinsic Capability of the Viscometer.

DIFFERENTIAL PRESSURE CAPILLARY VISCOMETER AS AN IN-LINE VISCOSITY DETECTOR

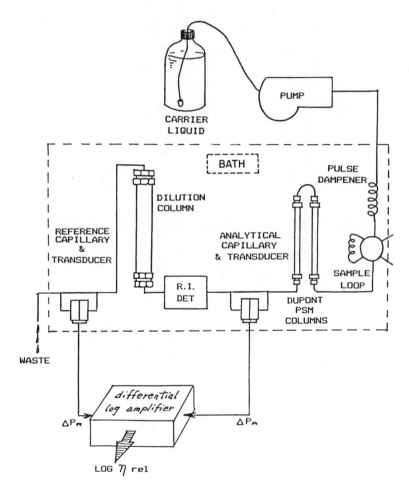

Figure 8. SEC Detector-Configuration 1.

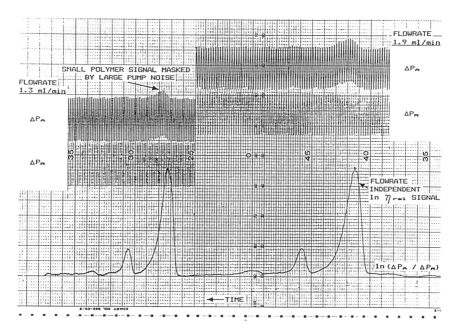

Figure 9. Flow Rate–Independent SEC–Viscosity Detection
Sample: polystyrene mixture (1.8M + 100K + 9K)
Solvent: THF, Concentration: 0.08%
2X Du Pont PSM–Bimodal columns 100 μl sample loop.

of the elution peaks remain the same at two flow rate levels is indicative of the viscosity detector providing the true viscosity information of the polymer sample. Typically, with an average good working LC pump, the SEC viscosity elution curves can be obtained with very little baseline noise, much better than the noise shown here. The usual SEC Sample loop size of 25 or 50 μl is normally adequate for the SEC viscosity analyses as well.

An alternative configuration for a SEC viscosity detector is shown in Figure 10. The depository column in the Figure 8 detector is replaced here by a considerably smaller delay volume. The delay volume element is nothing more than a coiled large i.d. tubing connecting the analytical and the reference capillary. The delay volume should be sufficiently large so that, when the separated polymer bands are eluting through the analytical capillary, only pure solvent is present in the reference capillary at the time. Typically, a delay volume of about 10 ml is sufficient for SEC systems where a set of two high performance SEC columns are used. The viscosity detector in Figure 10 is shown to have the concentration detector connected in the parallel arrangement. In operations, this viscosity detector will reset itself as the eluting polymer bands completely sweeps through the delay volume. Compared to the earlier detector configuration, this detector configuration offers a better signal-to-noise performance. The self-cleaning and reset feature is a considerable advantage. However, additional time is required with this detector configuration to flush out the delay volume after every sample.

Figure 11 illustrates an SEC separation of a sample of 3-component polystyrene mixture with the dual concentration and viscosity detectors of Figure 10. The top trace shows the concentration elution profile of the SEC separation as detected by a UV-photometer. The bottom trace records the same SEC separation, except with the viscometer signal from the log-amplifier output. The viscometer response is highly noise free and is shown here, as expected, being highly biased in favoring the detection of the early eluting high MW component. The last elution peak occurring before the flushing of the delay volume is caused by the impurity in the sample preparation. When the polymer sample is flushing through the delay volume and the reference capillary, a negative log-amplifier signal results as shown in the Figure. The flushing of the delay volume can be watched through the log-amplifier signal.

The high viscosity of some high MW samples is known to cause flow rate upsets as the sample passes through the SEC column frits. Such flow rate upsets often occur at the time of elution of the sample. While the flow rate upsets like this are likely to cause viscosity detection errors in most other SEC viscosity detectors, the signal of the present viscosity detector, however, will remain true, and unaffected by the high sample viscosity problem.

Placement of the reference capillary ahead of the SEC columns and the sample injection value is not an acceptable configuration for an SEC viscosity detector. Being at the high back pressure location, the flow rate noise sensed by the reference capillary would be out of phase with that sensed by the downstream analytical capillary. The compressibility of the column liquid volume under high pressure acts as an hydraulic capacitance causing the phase shifts of the flow noise between the two capillaries. The result is incomplete cancellation of flow rate fluctuations.

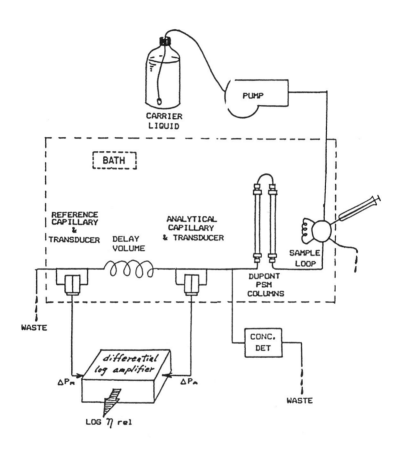

Figure 10. SEC Detector—Configuration 2.

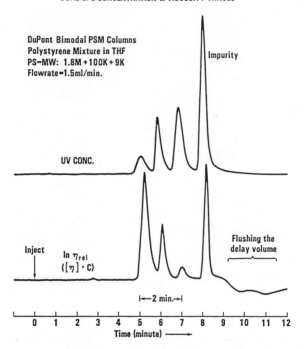

Figure 11. Typical SEC-Viscosity Analysis.
Sample: polystyrene mixture (1.8M + 100K + 9K)
solvent: THF, Concentration: 0.25% overall, mixture
ratio = 1:2:3.   2X Du Pont PSM-Bimodal columns 25 μl
sample loop, flow rate = 1.5 ml/mm.

SEC-MW and Universal Calibration

SEC is commonly known as a molecular-weight-distribution (MWD) technique. Strictly speaking, however, this is not quite true. SEC separates polymer molecules only according to their sizes rather than their MW. One finds that the SEC-MW calibration curves using the same column set can be quite different for different polymer-solvent systems, such as shown in Figure 12. Direct SEC-MW calibration is only possible when there are known MW standards available of the sample polymer type. This is only possible for a very limited number of polymer types shown in the Figure. For the majority of experimental polymers absolute GPC-MW calibration is simply impossible due to the lack of narrow MW standards.

One solution to the problem is by way of the universal calibration approach suggested by Benoit[11]. Since SEC separations are based mainly on the hydrodynamic volume of the polymer molecules, there should exist a universal SEC calibration curve when hydrodynamic volumes of molecules are used in the calibration plot. According to the basic theories in polymer science:

$$\text{Polymer hydrodynamic Volume} = [\eta]M \tag{14}$$

Therefore, the method of universal calibration would mean that a plot of the logarithm of $[\eta]M$ versus SEC elution volume V_R would behave like a master calibration plot that is unique for a particular SEC column set. Through the years, there have been numerous results published in support of the universal calibration practice in SEC. This is indeed true as illustrated by our data shown in Figures 12, 13 and 14. The very different MW and $[\eta]$ calibration curves of the four different polymer types shown in Figures 12 and 13 all converge to a single universal calibration curve based on the product of $M[\eta]$ in Figure 14.

As illustrated in Figure 15, the search for the MW calibration of an unknown polymer from the universal calibration curve requires an on-line SEC viscosity detector. Since narrow MW standards are not available for most commercial and experimental polymer samples, the determination of the polymer $[\eta]$ calibration would not be possible without an on-line SEC viscosity detector. Once the $[\eta]$ calibration for the unknown polymer is established, the polymer MW calibration can then be deduced from the universal calibration curve as indicated by the approaches shown in Figure 15.

SEC-[η] Calibration and Column Dispersion

The concentration of the polymer sample eluting from SEC column is sufficiently dilute that the recorded log-amplifier signal closely approximates the product $[\eta]xc$, in accord with Equation 12. Therefore, the $[\eta]$ value of the sample components can be determined by calculating the ratio between the viscosity and the concentration detector signals. This direct approach to $[\eta]$ calibration is illustrated in Figure 16, for two polystyrene samples: one is a broad MWD sample, the other is a mixture of three narrow standards. However, this $[\eta]$ calibration requires proper compensation for the effect of column dispersion and instrumental band-broadening. As shown in Figure 17, one finds the viscosity calibration resulting from a broad MWD sample normally tilts away from the true calibration curve. The extent of this

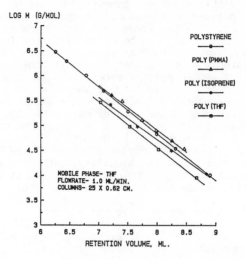

Figure 12. **Peak Position SEC-MW calibration**

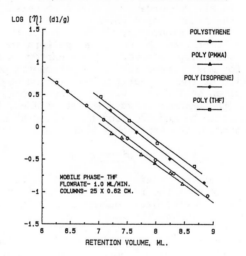

Figure 13. **Peak Position, SEC-Viscosity Calibration**

(Dupont Zorbax-PSM Trimodal Columns)

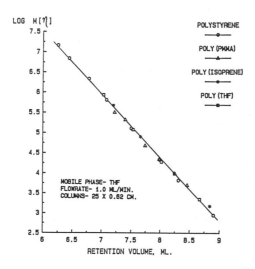

Figure 14. **Universal SEC calibration.**

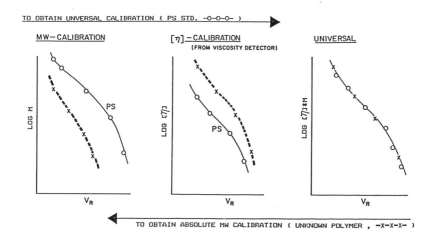

Figure 15. **MW Calibration Via Universal SEC Calibration.**

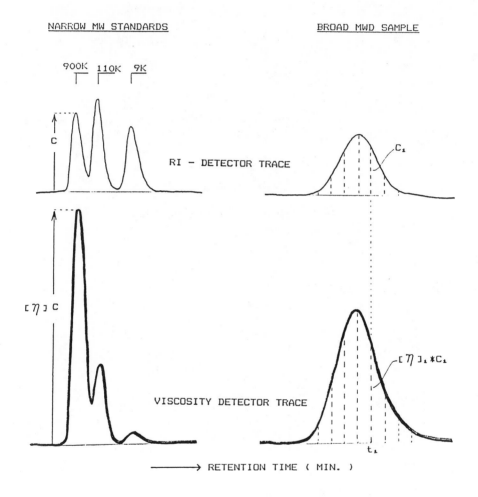

Figure 16. Typical SEC-Viscosity Analyses.

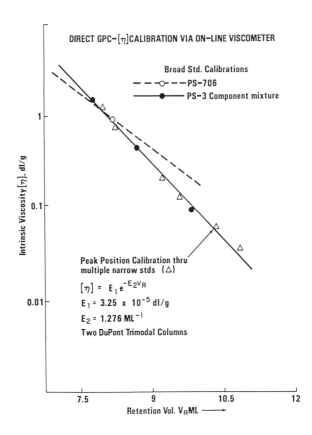

Figure 17. Direct SEC-[η] Calibration Via Viscosity Detector.

calibration curve mismatch is dependent on the extent of SEC column dispersion and the breadth of the sample MWD.[12,13] This effect of column dispersion on $[\eta]$ calibration could lead to serious errors in the SEC-MWD analysis using the universal calibration approach. Attempts have been made in our SEC-viscosity computer program to properly account for the SEC column dispersion effects. This, however, is another subject matter outside the scope of this paper.

Discussions

The preferred viscometer configuration for sensitivity and convenience is the design that uses the delay volume (Figure 5 and 10). This viscometer design has the unique advantage of being equally applicable as a batch viscometer as well as an SEC viscosity detector. Conversion between a viscometer or a detector involves the simple steps of changing to a different size sample loop and delay volume tubing. The viscometer hydraulic plumbing can be also easily modified to accommodate special sample handling, sample viscosity and shear-rate requirements.

The advent of new light scattering, viscosity, density, and photodiode detectors becoming available commercially could add a new dimension to SEC for studying polymer branching and copolymer composition problems. Proper computer software to treat the SEC analyses using multiple detectors is as important as the detector hardware development itself. SEC is a highly reproducible technique. The quantative aspects of the technique can be greatly expanded with the help of additional detectors and supporting software.

Literature Cited

1. Billingham, N.C., "Molar Mass Measurements in Polymer Science", John Wiley, New York, 1977.
2. Brandrup, J., and Immergut, E. H., eds., "Polymer Handbook," 2nd ed., Wiley, New York, 1975.
3. Quano, A. C., and Kaye, W., J. Polym. Sci., Part A-1, 12, 1151 (1974).
4. Meyerhoff, G., Makromol. Chem., 118, 265 (1968).
5. Quano, A.C., J. Polym. Sci., Part A-1, 10, 2169 (1972)
6. Lecacheux, D., Lesec, J., and Quivoron, C. J., Appl. Polym. Sci., 27, 4867 (1982).
7. Malihi, F. B., Kuo, C., Koehler, M. E., Provder, T., and Kah, A.F., Prepr. Org. Coat. Plast. Chem. Div. Am. Chem. Soc., 760 (1983)
8. Blair, D. E., "Capillary Viscometer," U.S. Patent #3,808,877 (1974).
9. Haney, M. A., "A New Differential Viscometer-Part One & Two", Am. Lab., March and April issues (1985); Haney, M.A.," Capillary Bridge Viscometer", U.S. Patent #4,463,598 (1984).
10. Abbott, S.D., Yau, W.W., "Differential Pressure Capillary Viscometer For Measuring Viscosity Independent of Flowrate and Temperature Fluctuations", U.S. Patent #4,578,990 and #4,627,271 (1986).
11. Benoit, H., Rempp, P., and Grubisic, Z., J. Polym. Sci., B5, 753 (1967).

12. Yau, W. W., Kirkland, J.J., Bly, D.D., "Modern Size-Exclusion Liquid Chromatography" Wiley (1979), Chap. 9, p. 306.
13. Cheng, R.S., And Bo, S.Q., ACS Symposium Series 245, "SEC Methodology and Characterization of Polymer and Related Materials", p. 125 (1984).

RECEIVED March 13, 1987

Chapter 6

An Experimental Evaluation of a New Commercial Viscometric Detector for Size Exclusion Chromatography

Using Linear and Branched Polymers

Mark G. Styring [1], John E. Armonas [2,3], and A. E. Hamielec [1]

[1]Department of Chemical Engineering, McMaster University, Hamilton, Ontario L8S 4L7, Canada
[2]Modchrom, Inc., 8666 Tyler Boulevard, Mentor, OH 44060

Herein are reported some findings on the application of a new type of continuous automatic viscometer, in parallel with a differential refractometer, as a detector system for SEC. A universal calibration is required for the instrument and two methods of construction are applicable. The first is the customary peak-position calibration using polymer standards of narrow molecular-weight distribution and the second uses a single broad standard of known M_w and M_n. The two types of calibration are shown to give nearly-identical values of molecular weight when used to process chromatograms obtained from various linear homopolymer standards of varying chemical composition. These values compare favourably with those quoted by the suppliers of the polymer standards. One of the more powerful features of this instrumentation, namely its potential for estimating accurate molecular weights of branched polymers, is demonstrated by analysis of a series of branched polyvinylacetates prepared by a conventional bulk, free-radical polymerisation procedure. The calculation of the degree of chain branching is discussed. Another particular feature of the viscometer detector, its ability to indicate the presence of low concentrations of high-molecular-weight impurity in polymer samples, is also shown.

Since its discovery in 1959 (1) and first application to synthetic organic polymers in 1964 (2), SEC has become the most widely used technique for routine characterisation of polymer molecular weights (MW) and molecular-weight distributions (MWD), as attested

---

[3]Correspondence should be addressed to this author.

0097–6156/87/0352–0104$06.00/0
© 1987 American Chemical Society

by the volume of publications which have appeared over the last twenty years. The basic technique and applications are well covered in books and reviews, eg. refs. 3-5. The major advantages of SEC over classical techniques (light scattering, osmometry, ultracentrifugation) are speed, simplicity of operation and reproducibility. A major drawback, however, in the case where a single detector (commonly a differential refractometer, DRI) is used to monitor column effluent, is the need for a calibration of the instrument, which reduces SEC to a secondary technique of polymer characterisation. Despite considerable experimental drawbacks and high costs, the idea of continuously monitoring the MW of the effluent as well as the concentration has proven attractive and powerful instrumentation has appeared over the past decade. Use of both an MW and a concentration detector in tandem provides data that can be used to make absolute MW calibrations.

The two techniques which have so far been applied to MW detection in SEC are low angle laser light scattering photometry (LALLSP) and viscometry (VISC). The earliest flow-through, light-scattering sample cell using a laser light source was described in 1966 by Cantow et al. (6). Pioneering work on a more useable system was done in the mid seventies by Ouano et al. (see eg. ref. 7) which led to the commercialisation of a reliable instrument, the KMX-6 and more recently the KMX-100, by the Chromatix Corporation. The earliest VISC detectors (8-10) in use sampled effluent discontinuously by successively charging a series of Ubbelohde-type viscometers. The first instrument which continuously measured viscosity, by monitoring the pressure drop across a capillary containing the column effluent, was described by Ouano (11) in 1972. The main problem with this design was a noisy signal due to extreme sensitivity of the pressure transducer in the flow cell to variations in solvent flow rate. Only recently has this problem been overcome. The Viscotek Model 100 instrument, which is a capillary-bridge viscometer, became commercialised in 1985 (12-14) and has a much improved signal-to-noise ratio. The purpose of the present work was an evaluation of this instrument. To this end a universal (hydrodynamic-volume) calibration (15) for our SEC system was constructed using well-characterised, narrow-as well as broad-MWD polystyrene (PS) and polymethylmethacrylate (PMMA) standards from commercial sources. The customary method for constructing a universal calibration is to determine MWs and intrinsic viscosities, $[\eta]$, for the polymers in question, analyze them on a suitable SEC system to obtain their retention volumes, V_R, then plot V_R against log $[\eta]M$. Separate determination of $[\eta]$ is not necessary when the Viscotek detector is employed in tandem with a suitable concentration detector, in our case a DRI. The Viscotek continuously monitors the pressure drop across a capillary tube which is proportional to the specific viscosity, η_{sp}, of the fluid flowing through it. Since the approximation

$$\eta_{sp} = [\eta] \, c \qquad (1)$$

is valid at the low polymer concentrations in SEC effluents and since we know c from the DRI detector response, then for any volume increment ΔV in a chromatogram.

$$[\eta] \, (\Delta V) = \eta_{sp} \, (\Delta V)/c \, (\Delta V) \qquad (2)$$

We can thus obtain $[\eta]$ for a slice of the chromatogram as narrow as we wish, or for the whole polymer. The only information required independently of our SEC system is the molecular weights of our calibration standards. This represents a significant advance over the previous situation in which $[\eta]$ had either to be determined in a separate experiment or through use of the Mark-Houwink parameters, K and a, if available from the literature.

Having obtained a suitable calibration for our system, the next step was to chromatograph a number of polymers of different chemical types having known MWs, namely polyvinylchloride (PVC), polysulphone, broad-MWD PMMA and both linear and branched polyvinylacetates (PVAc), on the VISC/DRI system, to compare the absolute MWs with those calculated from the instrumental data. Finally, some experiments were performed which demonstrate the utility of the VISC in detecting small amounts of high MW "impurity" in polymer samples.

Experimental

The majority of polymer samples examined in this work were of commercial origin. The suppliers' MW data were assumed to be accurate. Data pertaining to these polymers are given in Table 1.

Some PVAc samples examined in this study were synthesised in our laboratories at McMaster by a method which had been previously shown to yield products with extensive long-chain branching (LCB) (16). Freshly-distilled vinyl acetate monomer, to which was added 2,2'-azobis (2-methylpropionitrile) (AIBN) initiator at 1.02×10^{-4} mol. dm^{-3}, was pipetted into glass ampoules. These were plunged into liquid N_2 to freeze the monomer, attached to a vacuum line and evacuated for approx. 15 seconds, then sealed using a torch. Polymerisation was started by immersing the ampoules in an oil bath at $100 \pm 1°C$. Six sets of ampoules were removed at various intervals between 900 and 6300 seconds reaction time in the hope of obtaining polymer samples of varying MWs and degrees of branching In each case the ampoules were quenched in liquid nitrogen then broken open and a small quantity of 1-4-dioxane added to dissolve the polymer. After dissolution was complete, which took up to two days under ambient conditions, diethyl ether was added to the polymer solution in a flask immersed in iced water to bring about precipitation of the polymer. The product was finally melt evacuated at 130°C under high vacuum to constant weight.

An absolute value of M_w for each of these branched PVAcs was obtained from light-scattering measurements. In each case five polymer solutions were made up in tetrahydrofuran (THF) solvent and a Chromatix KMX-6 LALLSP instrument was employed to measure the intensity of light scattered from these solutions at 7° to the incident laser beam. A Chromatix KMX-16 laser differential refractometer was used to determine the refractive index increments, dn/dc, of the polymer solutions under ambient conditions.

Values of 0.0455 and 0.0528 $cm^3 g^{-1}$ were obtained respectively for the samples of lowest (900s) and highest (6300s) conversions. These are close to the value of 0.05 obtained for the same system by Atkinson and Dietz (17) who examined linear PVAcs, albeit using polarised light of wavelength $\lambda = 546$ nm as opposed to our laser light source of $\lambda = 633$ nm. Atkinson and Dietz also reported no significant variation of dn/dc with molar mass in the range $3 \times 10^4 - 1.5 \times 10^5$ g mol^{-1}, so it was decided to use the value 0.05 $cm^3 g^{-1}$ in each of our customary plots of Kc/R_θ against c to obtain M_w.

In order to demonstrate whether or not the PVAcs we had synthesised were branched it was decided to perform some measurements using an SEC instrument with a single DRI detector. The instrumentation comprised a Waters Associates ALC-150 automated liquid chromatograph fitted with five Toyo Soda H Gel columns designated 1000, 2500, 3000, 4000 and 6000. THF was pumped at 1.0 cm^3/min through the system which was kept at 30°C. The columns were calibrated with a kit of ten narrow-MWD linear PS standards from the Toyo Soda Co. Our PVAcs were then chromatographed on the same instrument (sample concentrations $\simeq 1$ mg/ml). The Mark-Houwink constants for linear polystyrene in THF at 25°C (15), $K_1 = 1.50 \times 10^{-2}$ $cm^3 g^{-1}$, $a_1 = 0.70$ and for linear PVAc in THF at 35°C (17), $K_2 = 1.56 \times 10^{-2}$ $cm^3 g^{-1}$, $a_2 = 0.708$ were employed to obtain estimates of MW for our PVAcs based on their peak retention volumes. This was performed assuming validity of the universal ($[\eta]M$) calibration procedure, ie. at any V_R

$$[\eta]_1 M_1 = [\eta]_2 M_2 \qquad (3)$$

TABLE 1

Data Pertaining to Linear Polymers of Commercial Origin Used in This Study

Sample Identification		Supplier and Lot number	M_w /10^3gmol^{-1}	M_n /10^3gmol^{-1}	c* mg/ml
PS Narrow	2070	PC – 61222	2.07	1.79	1.73
"	3970	PC – 61110	3.97	3.57	1.69
"	8770	PC – 80314	8.77	–	1.71
"	18100	PC – 41220	18.1	15.1	1.66
"	46500	PC – 30908	46.5	46.3	0.81
"	114K	GY – S – 8	114	108	0.33
"	168K	PC – 30126	168	164	0.42
"	262K	PC – 50124	262	258	0.47
"	402K	PC – 00507	402	395	0.32
"	599K	PC – 30121	599	585	0.42
"	940K	PC – 80323	940	925	0.27
"	1560K	PC – 50329	1560	1495	0.41
"	1894K	PC – 50724	1894	1790	0.36
"	2817K	WA – 14B	2817	2300	0.34
PS Broad Dow 1683		Dow 1683	250	100	2.29
"	321K	SPP	321	84.6	2.60
PMMA Narrow 7670		PC – PM5 – 1	8.01	6.96	2.0
"	27K	PC – PM5 – 2	27.0	24.3	0.46
"	49K	PC – PM5 – 3	49.0	40.0	0.81
"	107K	PC – PM5 – 5	107	107	0.54
"	240K	PC – PM3 – 5	240	266	0.30
"	330K	PC – PM5 – 9	330	295	0.63
"	400K	PC – PM5 – 10	400	360	0.22
"	840K	PC – PM5 – 12	840	750	0.30
PMMA Broad 33K		PS – 06	33.3	13.7	2.62
"	93K	PS – 09	93.3	46.4	1.54
"	490K	PS – 08	490	119.2	1.59
"	699K	PS – 16215	699	213	0.49
Polysulphone		SPP – 01	67.0	20.4	3.25
PVC	77K	SPP – 02	77.3	39.6	1.47
"	122K	SPP – 03	122	57.3	1.85
"	193K	SPP – 04	194	86.3	1.67
PVAc	125K	SPP – 05	125	52.7	2.75
"	195K	SPP – 06	195	63.6	2.06
"	237K	SPP – 07	237	89.9	2.01

Note:

c*   denotes concentration of sample injected into VISC/DRI SEC system.

Suppliers:

PC   denotes Pressure Chemical Co., Pittsburgh, Pa.

GY   denotes Goodyear Tyre and Rubber Co., Akron, Oh.

WA   denotes Waters Associates, Milford, Ma.

PS   denotes Polysciences, Warrington, Pa

SPP   denotes Scientific Polymer Products, Ontario, N.Y.

$$\log M_2 = \log (K_1/K_2) + ((a_1 + 1) \log M_1) / (a_2 + 1) \tag{4}$$

The instrument which formed the basis of this investigation, an SEC with a dual VISC/DRI detector system, is now briefly described, together with basic operational information. Degassed UV grade THF was pumped through the system at 1.5 cm^3/min by a Waters Associates Model 6000 A pump. Two Modchrom mixed-bed columns, each 7.8 mm $\times$ 30 cm, packed with porous particles of nominal diameter 5 μm (similar to μ-Styragel), were maintained at 50°C in a Waters TCM column oven. Samples were applied by means of a Rheodyne 7125 six-port injector equipped with a 100 μl sample loop. The two detectors, fitted in parallel, were a Knauer LED (DRI), operating at 21-22°C and a Viscotek Model 100 (VISC) operating at 30°C. Data acquisition and analysis were performed on an IBM PC/XT computer using an ASYST – UNICAL 2.04 software system as supplied with the Viscotek instrument.

The experimental plan for this SEC system was as follows. First, calibration was achieved using a single broad standard PS of known M_n, M_w and [η] ie. Dow 1683. Knowledge of [η] allows one to set the instrument parameters for the VISC, the most important being the differential pressure transducer (DPT) sensitivity. This was accomplished by setting an initial value for the instrument parameters and establishing a calibration using the Dow 1683 chromatograms by supplying the known values of M_w and M_n. The same chromatogram was then analysed using the newly-established calibration to give back M_w, M_n and [η]. This latter step was repeated as often as necessary, each time entering a modified value of DPT sensitivity, until the correct value of [η] was obtained. (N.B in our work the optimum value of DPT sens. was 1.30. This gave [η] = 0.841 dl/g which is exactly the value quoted by the supplier). Next, peak-position calibrations were established using linear, narrow-MWD PS and PMMA standards. A comparison of the broad PS, narrow PS and narrow PMMA calibrations was then made by comparing values of MW calculated using each one from chromatograms of selected known samples.

Secondly, a series of linear, well-characterised polymers (PVCs, PMMAs, PVAcs and polysulphone) were analyzed and their MWs by SEC compared with those quoted by the supplier.

Thirdly, we examined the branched PVAcs, mentioned earlier. A comparison was made of the three sets of MW data calculated from the LALLS, the SEC-DRI system and the SEC-dual VISC/DRI system.

Finally, we sought to demonstrate the efficiency of the VISC detector in measuring small quantities of high-MW impurities in polymer samples. This was done by analyzing solutions of the three PS blends made from PS Narrow 2817 K and PS Narrow 168 K with small, successively increasing weight fractions of the former.

Results and Discussion

a) Calibration. Figure 1 shows the (linear) calibration established using the Dow 1683 broad PS standard. Figure 2 shows the calibration established using 14 narrow-MWD PS standards together on the same plot with data for the 8 narrow-MWD PMMAs. In constructing these two calibrations, values of [η] and peak V_R were determined by the instrument. The MWs chosen were M_ws, from the suppliers (Table 1). Clearly the PMMA data fit very well on the PS curve in fig. 2, which is intuitively satisfying.

Table 2 gives a comparison of the values of M_w and M_n calculated from the VISC/DRI chromatrograms of three standards using the three calibrations described above. The two PS calibrations in particular give excellent agreement with the Dow 1683 data. M_w values are within 10% of those quoted for both PS 321K and PMMA 33K although the M_n values are in somewhat greater disagreement. This discrepancy is discussed later.

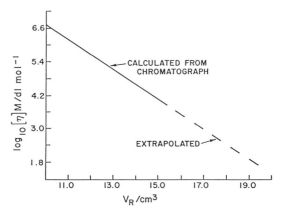

Figure 1   Universal calibration curve established using a single broad-MWD PS standard of known M_w and M_n.

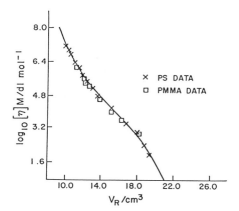

Figure 2   Universal calibration curves established using a series of narrow-MWD PS and PMMA standards.

TABLE 2

Comparison of MW Data from Supplies with Those Calculated from
the Three Types of Calibration Established for the VSIC/DRI SEC

Sample	M_w /10^3gmol^{-1}	M_n /10^3gmol^{-1}	Calibration
Dow 1683	251.9	101.3	PS broad
broad	248.5	97.8	PS narrow
standard	247.0	118.4	PMMA narrow
	(250.0)	(100.0)	(Suppliers' value)
PS 321K	295.6	111.3	PS broad
broad	332.1	128.1	PS narrow
standard	301.2	129.4	PMMA narrow
	(321.0)	(84.6)	(Suppliers' value)
PMMA 33K	36.8	15.7	PS broad
narrow	36.7	19.3	PS narrow
standard	38.0	20.9	PMMA narrow
	(33.3)	(13.7)	(Suppliers' value)

Figure 3 is illustrative of the type of errors which could be incurred if one attempted to use a single narrow-MWD standard to calibrate an SEC instrument. Fig. 3(a) is an intrinsic-viscosity plot calculated from the DRI and VISC responses for PS Narrow 114K. Plotted on the same graph is a portion of the [η] versus V_R curve constructed from the [η] and peak V_R values of the narrow-MWD PS standards given by the VISC/DRI system, which we call the "ideal" curve (see fig. 4). Fig. 3(b) is a similar plot for PS Narrow 940K. Although the value of [η] around the peak V_R is close to the ideal value, departure from ideality occurs very rapidly as one moves away from the peak in both cases. This is a result of peak-broadening effects which are most marked for narrow-MWD polymers. Even though the chromatograms were corrected for broadening, an entirely satisfactory algorithm for dispersion correction has yet to be devised owing to the complexity of the transport phenomena within SEC columns giving rise to the broadened peaks. Fig. 3(c) is the intrinsic-viscosity plot for the Dow 1683 broad PS standard. The curve is the actual plot and the data points are taken from the "ideal" curve. Agreement between the two sets of data is very good, which is validification of the single-broad-standard calibration method. Owing to the actual width of the MWD in such standards, the errors incurred in applying peak-broadening corrections are relatively much smaller than is the case for narrow-MWD polymers.

b) Linear Polymer Standards. The suppliers' values of M_w and M_n for linear PVC, PMMA, PVAc and polysulphone samples are given in Table 3, together with those calculated from VISC/DRI chromatographic data using the narrow PS calibration. Once again the calculated M_w values are generally in very good agreement with the quoted ones, whilst the M_n values are somewhat at variance. There are two possible reasons for the latter. One is the reliability of the quoted M_n values. Some of these were themselves calculated from other SEC data with the concomitant pitfalls. The second and possibly more important factor is the fact that the current Viscotek-supplied software uses

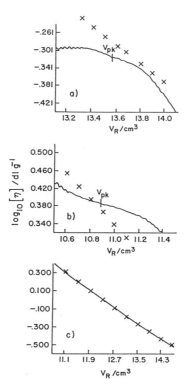

Figure 3   Intrinsic-viscosity plots for the polystyrene standards a) PS Narrow 114K, b) PS Narrow 940K and c) Dow 1683 (broad).

constant values of σ (the Gaussian broadening parameter) and τ (the skewing factor) in correcting peaks for broadening. It is well known (see, eg. ref. 4), however, that these parameters do vary with V_R. This is illustrated in figure 5 for our SEC system where values of σ and τ, as calculated from the VISC/DRI chromatograms, for eleven of the narrow PS standards are plotted as a function of peak V_R. The peak-broadening correction routine in the ASYST-UNICAL 2.04 software accepts only a single value of either of these parameters and recommends the use of those obtained from the narrowest standard; in our case PS Narrow 262K.

TABLE 3

Comparison of MW Data from Suppliers with Those Calculated from the Narrow-PS Calibration for Linear PVCs, PMMAs, PVAcs and Polysulphone

Sample		$M_w/10^3 gmol^{-1}$		$M_n/10^3 gmol^{-1}$	
PVC	77K	71	(77)	33	(40)
	122K	82	(122)	45	(57)
	193K	175	(193)	81	(86)
PMMA	33K	37	(33)	19	(14)
	93K	102	(93)	65	(46)
	490K	521	(490)	233	(119)
	699K	695	(699)	264	(213)
PVAc	124K	125	(124)	51	(64)
	194K	218	(194)	91	(53)
	237K	242	(237)	72	(90)
Polysulphone		67	(67)	32	(20)

Figures in brackets refer to suppliers' quoted values.

c) Branched PVAcs. Table 4 gives the values of M_w for the six samples calculated from the LALLSP data, the Waters ALC-150 chromatographic data (using Mark-Houwink parameters for linear PS and linear PVAc) and from the VISC/DRI SEC system. Considering the LALLSP data alone the increase in M_w with conversion is in accord with previously-published data (16,18) and is bound up with the kinetics of bulk, free-radical VAc polymerisation (18). Comparison of the M_p values (MWs calculated at the peak V_R) from the ALC-150 instrument with the M_ws from LALLSP shows a consistent, sizeable underestimate in each case for the chromatographic data. That the use of Mark-Houwink constants for linear PVAc gives rise to such underestimates is good evidence that the polymers are in fact branched. Branched polymers have more compact conformations and hence lower intrinsic viscosities than their linear analogues. Substitution of a value of $[\eta]_2$ which assumes linearity in equation 3 will then lead to too low a value of M_2 if the polymer happens to be branched. The values of M_w calculated from the VISC/DRI data, where $[\eta]$ is measured directly for each whole-polymer sample, branched or unbranched, compare much more favourably with the LALLSP values.

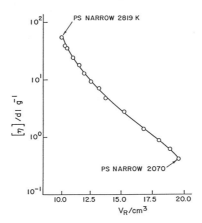

Figure 4   "Ideal" intrinsic viscosity vs. peak V_R plot, constructed from narrow-MWD PS standard data.

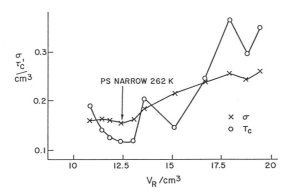

Figure 5   The variation of the peak-broadening parameters, σ and τ, against V_R for eleven narrow-MWD PS standards.

TABLE 4

Value of M_w for branched PVAcs calculated using one absolute technique (LALLSP) and two SEC instruments (DRI and dual VISC/DRI)

PVAc Identification	M_n (Waters A(C − 150))		M_w(LALLSP)	M_w (VISC/DRI)	
			MWs/10³ g mol⁻¹		
900	270	(− 18%)	328	402	(+ 23%)
1900	270	(− 34%)	411	434	(+ 1%)
3600	279	(− 31%)	406	422	(+ 4%)
4500	279	(− 42%)	482	436	(−10%)
5400	321	(− 33%)	479	456	(− 4%)
6300	396	(−37%)	628	539	(− 14%)

Percentages in brackets refer to discrepancy between chromatographic and absolute (LALLSP) M_w data.

No attempt was made to calculate actual degrees of branching for those PVAcs in the present work. The computational scheme for so doing is quite straightforward, but the exact meaning of such estimates is open to question owing to the many assumptions, valid and invalid, required to derive a quantitative estimate of branching from SEC and viscometric data.

The experimental quantity from which information concerning branching is derived is g', the ratio of intrinsic viscosities of branched (subscript br) and linear (subscript l) isomers;

$$g' = [\eta]_{br} / [\eta]_l \qquad (5)$$

We need further to infer from g' an estimate of g_0 the ratio of unperturbed radii of branched and linear isomers:

$$g_0 = < r_g^2 >_{br}^{1/2} / < r_g^2 >_l^{1/2} \qquad (6)$$

Relations between g' and g_0 are semi empirical and approximate (19,20). It is assumed that g' is independent of solvent conditions and that a theta solvent for a linear polymer is also a theta solvent for its branched analogues. Neither of these assumptions is well founded (19). In practical applications, exponential relations between g' and g_0 of the form

$$g' = g_0^k \qquad (7)$$

have been proposed. For example, a well-studied branched polymer is low-density polyethylene (LDPE), for which k = 1.2 has proven satisfactory (21-23). The number, n, of branches per macromolecule is obtained through one of the Zimm-Stockmayer relationships (24), an example of which is given below:

$$g_0 = \left[\left(1 + n/7 \right)^{1/2} + 4n/9\pi \right]^{-1/2} \tag{8}$$

In outline, a method of calculating n using the VISC/DRI system is as follows. First, one should establish a universal calibration for the instrument. Second, obtain chromatograms for the branched polymer, together with a value of $[\eta]_{br}$. M_{br} is calculated directly from the universal calibration. An estimate of $[\eta]_l$ is then obtained using a suitable Mark-Houwink relationship and we now have sufficient information to calculate g' and hence n through equations 5-8. The great advantage of the VISC/DRI system over earlier methods of determining branching by SEC and off-line viscometric measurements is that it is possible to measure $[\eta]_{br}$ across the chromatograms in slices as small as we desire. This means that we could examine how branching varies with MW across a given sample, as well as obtaining the average value of n for the whole polymer.

d) Detection of Small Amounts of High-MW Impurities. The dual chromatograms for the three blends of PS Narrow 2817 K and PS Narrow 168 K are shown in fig. 6. Estimates of the amount of high-MW material calculated separately from the DRI and from the VISC detector responses are shown alongside each dual chromatogram, together with the amount actually weighed in to the sample vial. Clearly, there is a substantial difference between the amounts calculated from the detector responses and those obtained by direct weighing. A number of reasons for this are possible, the most important being the method of calculation. For a DRI response, at any V_R, the chromatogram height $H_{DRI}(V)$ is proportional to the concentration c(V). Hence, for the volume interval ΔV

$$c_{DRI} (\Delta V) = \left(\sum_{\Delta V} H_{DRI} (V) / \sum_{all\ V} H_{DRI} (V) \right) \times Total\ c \tag{9}$$

For a VISC response $H_{VISC} (V)$ is proportional to the specific viscosity, $\eta_{sp} (V)$, i.e. is proportional to $[\eta] (V) c(V)$. Hence

$$c_{VISC} (\Delta V) = \left(\sum_{\Delta V} (H_{VISC} (V) / [\eta] (V)) / \sum_{all\ V} (H_{VISC} (V) / [\eta] (V)) \right) \times Total\ c \tag{10}$$

Values of $[\eta] (V)$ were obtained directly from the curve shown in fig. 4. All our calculations were performed manually by taking approximately 40 raw chromatogram heights from across the entire sample. Greater accuracy would be achieved using a computer with a more accurate integration routine (based, eg., on Simpson Rule).

Whatever numerical errors are incurred in calculating percentage concentrations from detector responses, it is clear from fig. 6 that the viscometer detector gives a much better qualitative indication of the presence of high-MW species in a sample than the DRI. This is particularly noticeable in fig. 6a where the presence of 1.5% high-MW polymer is barely discernible from noise in the DRI trace, whilst a substantial peak is obtained from the VISC.

It is noteworthy that the VISC detector would not be particularly useful in the detection of microgel, i.e. high-MW cross-linked impurity in polymer samples. Microgels have very low viscosities, despite their high MWs, owing to their compact, spherical conformations. We can illustrate this with reference to a specific example. A series of polystyrene-co-divinylbenzene microgels containing approximately 5% cross-linking agent were prepared by Booth et al. (25). Subsequent intrinsic-viscosity determinations

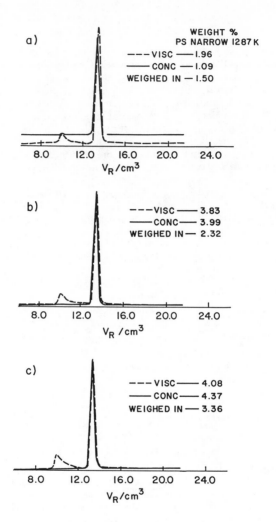

Figure 6   Dual chromatograms obtained from polystyrene blends of PS Narrow 2817K
and PS Narrow 168K.

revealed that the microgels had intrinsic viscosities which were approximately one-thirtieth the values obtained for linear polystyrenes of equivalent MW.

Summary

The present work has been concerned with the evaluation of an SEC system with dual detectors; a DRI coupled with a new type of differential viscometer. The system was calibrated with both narrow-and broad-MWD polymer standards and the accuracy of each type of universal calibration checked by chromatographing various linear, well-characterised polymers. Good agreement was obtained between the quoted M_w values and those calculated from the chromatograms. M_n values were at somewhat greater variance. Several possible reasons for this were discussed. Some branched PVAcs were synthesised and characterised by an absolute technique, LALLSP. Values of M_w obtained from the VISC/DRI system were in very good agreement with the absolute data, which gives further validification to the universal-calibration method of Benoit, et al (15) and illustrates the power of this new SEC system as a rapid, reliable technique of characterising branched polymers.

Finally, we have demonstrated the utility of the viscometer in detecting small amounts of high-MW impurity in polymer samples.

Acknowledgements

M.G.S. and A.E.H. are grateful to the Natural Sciences and Engineering Research Council of Canada for financial support of this work.

M.G.S. is particularly indebted to the Science and Engineering Research Council of the U.K. for the award of an overseas Postdoctoral Fellowship.

The authors would like to thank Miss Lisa Lee for her care in obtaining some of the experimental data in this investigation.

References

1.   Porath, J. and Flodin, P. Nature, 1959, 183, 1657.
2.   Moore, J.C. J. Polym. Sci. Part A, 1964, 2, 835.
3.   Hagnauer, G.L. Anal. Chem., 1982, 54, 265R.
4.   Yau, W.W.; Kirkland, J.J. and Bly, D.D. "Modern Size Exclusion Liquid Chromatography. Practice of Gel Permeation and Gel Filtration Chromatography"; Wiley: New York, 1979.
5.   "Steric Exclusion Liquid Chromatography of Polymers"; Chromatog. Sci. Series, 25; Janca, J.; Ed; Dekker: New York, 1984.
6.   Cantow, H-J; Siefert, E. and Kuhn, R. Chem. Eng. Tech., 1966, 38, 1032.
7.   Ouano, A.C. and Kaye, W. J. Polym. Sci. Part A1, 1974, 12, 1151.
8.   Meyerhoff, G. Makromol. Chem., 1968, 118, 265.
9.   Goedhart, D. and Opschoor, A. J. Polym. Sci. Part A2, 1970, 8, 1227.
10. Meyerhoff, G. Sepn. Sci., 1971, 6, 239.
11. Ouano, A.C. J. Polym. Sci. Part A1, 1972, 10, 2169.
12. Haney, M.A. U.S. Patent 4,463,598, 1984.
13. ibid. Am. Lab., 1985, 17, 41.
14. ibid. p. 116.
15. Benoit, H; Grubisic, Z and Rempp. R. J. Polym. Sci. Part B, 1967, 5, 753.
16. Hamielec, A.E.; Ouano, A.C. and Nebenzahl, L.L. J. Liquid Chromatog., 1978, 1, 527.
17. Atkinson, C.M.L. and Dietz, R. Eur. Polym. J., 1979, 15, 21.
18. Graessley, W.W.; Hartung, R.P. and Uy, W.C. J. Polym Sci. Part A2, 1969, 1, 1919.

19.   Small, P. Adv. Polym. Sci., 1975, 18, 1.
20.   Dietz, R. and Francis, M.A. Polymer, 1979, 20, 450.
21.   Hama, T.; Yamagushi, K. and Suzuki, T. Makromol. Chem., 1972, 155, 283.
22.   Völker, H. and Luig, F.J. Angew. Makromol. Chem., 1970, 12, 43.
23.   Lecacheux, D.; Lesec, J. and Quivoron, C. J.Appl. Polym. Sci., 1982, 27, 4867.
24.   Zimm, B.H. and Stockmayer, W.H. J. Chem. Phys., 1949, 17, 1301.
25.   Booth, C.; Forget, J-L.; Georgii, I., Li, W.S. and Price, C. Eur. Polym. J., 1980, 16, 255.

RECEIVED August 18, 1987

Chapter 7

Gas Permeation Chromatography–Viscometry of Polystyrene Standards in Tetrahydrofuran

M. A. Haney [1], John E. Armonas[2], and L. Rosen [3]

[1]Viscotek Corporation, 1030 Russell Drive, Porter, TX 77365
[2]Modchrom, Inc., 8666 Tyler Boulevard, Mentor, OH 44060
[3]Pressure Chemical Company, 3419 Smallman, Pittsburgh, PA 15201

Studies of polystyrene standards in THF solvent are not uncommon. However, bothersome discrepancies still exist in the literature and in practice. For example, the published Mark-Houwink parameters are in wide disagreement (1). The purpose of this work is to examine a large number of PS standards from multiple suppliers, covering a wide range of molecular weights. The intrinsic viscosities and GPC retention volumes have been measured and used independently to correlate and crosscheck the molecular weights provided by the suppliers.

EXPERIMENTAL

The size exclusion chromatography for this study was done in the routine manner execept for the inclusion of an on-line viscosity detector called a Differential Viscometer (3) (Viscotek Corp., Porter, Texas, USA). This instrument together with an RI concentration detector permits the calculation of intrinsic viscosities across the chromatogram. An IBM PC data system with software is also provided (5). The software acquires data from both detectors, and performs calculations of intrinsic viscosity and molecular weight distributions using the Universal Calibration Method.

Columns:	Two Modchrom Mixed Bed Styrene-divinylbenzene 5 micron Gel
Solvent:	THF (stabilized)
Flow rate:	1.5 ml/min
Temperature:	30 C
Pump:	Knauer Model 64
Injector:	Rheodyne M 7125, fixed 50 ul loop

0097–6156/87/0352–0119$06.00/0
© 1987 American Chemical Society

Sample Conc   0.8 % for M < 1000
                     0.25 % for 1000 < M < 300,000
                     0.08 % for M > 300,000
Conc Detector: Knauer Differential Refractometer
                     Model LED
Viscosity Detector: Viscotek Differential Viscometer
                     Model 100
Data Acquisition: Viscotek Unical DS

RESULTS AND DISCUSSION

Data are shown in Table I.   Intrinsic viscosities were
determined in separate experiments by the batch differential viscometry method (2) as shown in column 2 and by
the on-line GPC differential viscometry method (3) as
shown in column 3.   GPC peak retention volumes are shown
in column 4.   Column 5 lists the moleculare weights provided for these standards by the suppliers.   These were
taken to be the Mw values from light-scattering measurements, where available.

Figure 1 shows the logarithmic plot of intrinsic viscosities vs M(supplier).   The plot has a distinct "break"
at a molecular weight of about 10,000.   This type of break
in the plot has been observed for many polymers(4).   Above
the break the Mark-Houwink constants are similar, but
distinct from those previously reported for polystyrene
in THF (1).   Below the break the Mark-Houwink constants
are drastically different.   The molecular weights computed
from the experimental intrinsic viscosities using the
derived Mark-Houwink constants are shown in column 6 of
Table I.

Figure 2 shows a conventional GPC calibration curve for
the standards, that is, a plot of GPC retention vs log M.
These data were fitted to a 4th order polynomial and the
molecular weights of all the standards recomputed from
this fit using the measured retention volumes.   These
values are shown in column 7 of Table I.

In most cases, the derived molecular weights in columns 6
and 7 are fairly consistent with the supplier moleculare
weights in column 5.   The "best" values in column 8 are
the averages of the values in columns 5-7 for those cases.
In other cases, the derived values appear to be distinctly
different from the supplier values, in which case the
"best" values are taken to be the average of columns 6-7
It should be noted that the data in columns 6-7 cannot be
more accurate or better than the data in column 5 on a
collective basis, because the former has been derived
from the latter.   However, on an individual basis, the
data in columns 6-7 should be more accurate, because it is

Table I. Data on Narrow Polystyrene Standards

Supplier-Lot no.	IV Batch (dl/g)	IV On-line (dl/g)	Ret. Vol. (ml)	M Supplier (×1000)	M Mark-H (×1000)	M Ret.Vol (×1000)	M "Best" (×1000)
GY-S-11	--	.0255	20.82	.666	.557	.548	.553
GY-S-14	--	.0265	20.77	.517	.609	.583	.596
PC-50822	.031	.0277	20.51	.800	.752	.792	.781
PC-11103	.029	.0277	20.55	.826	.710	.756	.764
GY-S-13	.035	.0330	20.35	.910	1.09	.948	.982
PC-50211	.039	.0382	19.83	1.35	1.47	1.64	1.49
PC-61222	.044	.0428	19.49	2.00	1.93	2.28	2.07
WA-61222	--	--	--	1.85 (2.35 on bottle)			
GY-S-9	.044	.0432	19.67	1.94	1.95	1.92	1.94
PC-61110	.058	.059	18.85	4.00	3.88	4.02	3.97
PC-30525	.060	.062	18.77	4.00	4.28	4.30	4.19
GY-S-10	.069	.066	18.48	5.48	5.42	5.42	5.45
PC-50828	--	.073	18.40	6.00	6.52	5.77	6.10
GY-S-12	.076	.075	18.14	7.82	7.05	7.03	7.04
PC-80314	.085	.085	17.96	9.00	9.30	8.01	8.77
WA-80314	--	--	--	8.50 (15.0 on bottle)			
GY-S-3	.107	.105	17.38	12.2	12.6	12.0	12.3
PC-30420	.119	.119	17.22	13.8	14.8	13.3	14.0
PC-41220	.139	.135	16.77	18.5	18.0	17.7	18.1
GY-S-5	.152	.150	16.54	19.4	20.8	20.4	20.6
PC-30811	.170	.170	16.20	23.0	24.4	25.1	24.2
PC-80317	.202	.201	15.87	35.0	31.0	30.5	30.8
PC-30908	.264	.265	15.15	47.5	45.4	46.5	46.5

Continued on next page.

Table I. Continued

Supplier-Lot no.	IV (dl/g) Batch	IV On-line	(ml) Ret. Vol.	M Supplier	M (×1000) Mark-H	M Ret.Vol	M "Best"
GY-S-7	.251	.286	14.91	47.5	46.0	53.5	49.0
GY-S-6	.373	.371	14.19	79.0	73.4	82.4	78.3
PC-70111	.443	.450	13.96	93.0	95.0	95.1	94.4
WA-70111	---	---	---	100 (110 on bottle)			
GY-S-8	.675	.490	13.56	112	108	122	114
PC-30126	.945	.687	13.13	169	171	164	168
PC-50124	1.25	.962	12.50	254	275	258	262
PC-00507	1.26	1.27	11.96	---	410	394	402
PC-3B	---	1.28	11.96	392	412	394	400
WA-3B		---	---	470 (on bottle)			
GY-S-2	---	1.50	11.50	612	520	582	571
PC-30121	1.69	1.66	11.47	591	608	598	599
PC-80323	2.28	2.31	10.98	929	949	942	940
GY-S-1	---	3.04	10.66	1460	1403	1294	1386
PC-50329	3.22	3.37	10.52	1610	1575	1496	1560
PC-30622	3.71	3.82	10.32	1760	1892	1852	1872
PC-50724	3.85	3.89	10.32	1860	1970	1852	1894
WA-14B	4.95	4.85	9.89	2700	2745	3007	2817

(on bottle)

NOTES: PC designates supplier as Pressure Chemical Co., Pittsburgh, Pa.
GY designates supplier as Goodyear Tire & Rubber Co., Akron, O.
WA designates supplier as Waters Associates, Milford, Mass.

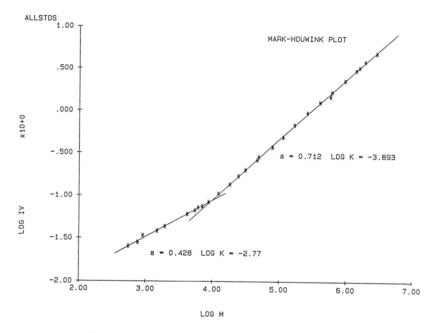

Figure 1.   Plot of log IV vs Log M (Supplier)

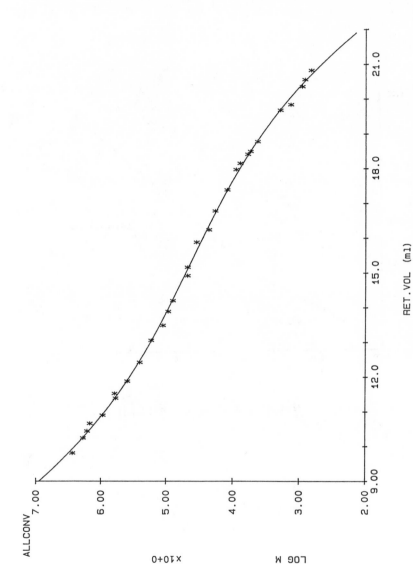

Figure 2. Conventional Calibration of Data from Table I using Supplier Molecular Weights

derived from additional measurements and using a larger data base than was available to the suppliers.

The "best" molecular weights were then combined with the measured intrinsic viscosities to generate a Universal Calibration Curve shown in Figure 3. This curve was used to analyze several broad distribution polystyrene standards using the universal calibration method and measuring the intrinsic viscosity distribution of the broad standards across the chromatograms with the Differential Viscometer Detector. This method generates molecular weight distribution and Mark-Houwink parameters a and k directly. The data are shown in Table II.

Initial runs on Dow 1683 and NBS 706 showed that the experimental values of Mn and Mw seemed to be too high compared to the standard values. This was felt to result from the use of concentrations too high in the narrow standards. One of the standards (168 K) was run at various concentrations to see the effect of concentration on measured retention volume(Table III.). It can be seen that at this molecular weight, the concentration of sample should have been less than 0.05 % to prevent concentration effects. In fact, the samples in this concentration range were about 0.25 % , which leads to a correction of about .08 ml to the narrow standards. This correction was applied to the retention volumes of the narrow standards above 100K and the broad standards were recomputed from the corrected universal calibration curve. These results are shown in Table IIb. Now the values for Dow 1683 and NBS 706 are comparable to the standard values . Also shown in Table IIb are values obtained for two experimental polystyrene standards, Dow 685 and SPP 039C.

There is some scatter in the computed Mark-Houwink constants from the broad polystyrene standards of Table IIb. However, the average of the four samples is very close to that found from the set of narrow standards.

CONCLUSIONS

1. The intrinsic viscosity plot for polystyrenes shows a distinct break at about 10,000 daltons.
   Above this value the Mark-Houwink parameters are:
   $$a = .712 , \quad K = 1.28 \times 10^{-4} .$$
   Below 10,000 the Mark-Houwink parameters are:
   $$a = .428 , \quad K = 1.71 \times 10^{-3} .$$

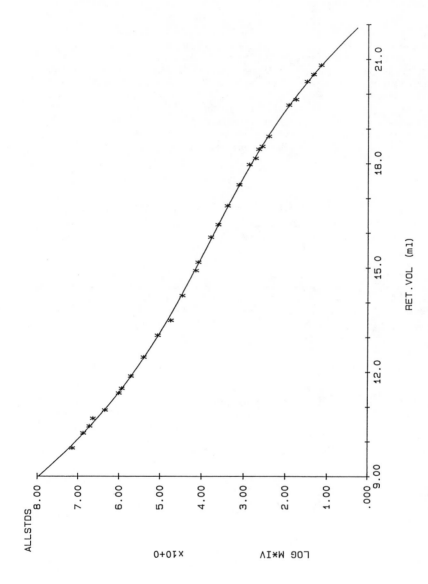

Figure 3.   Universal of Data from Table I using "Best" Molecular Weights

Table II. Data on Polystyrene Broad Standards

Method: Universal Calibration

a. using original calibration:

	M_n	M_w	M_z	IV	a	$-\log K$
DOW 1683	116 (100)*	279 (250)	484 (430)	.849	.703	3.865
NBS 706	139 (122)	318 (275)	504	.879	.693	3.842

b. after correcting calibration for concentration effects:

	M_n	M_w	M_z	IV	a	$-\log K$
DOW 1683	103 (100)	246 (250)	419	.841	.740	4.035
NBS 706	114 (122)	272 (275)	428	.869	.736	4.030
SPP 039c	134 (84.6)	335 (321)	755	.959	.711	3.901
DOW 685	167 (110)	315 (285)	517	.967	.711	3.894
				avg=	.722	3.965

NOTES: DOW designates Dow Chemical Co., Midland, Mich. as supplier.
NBS designates National Bureau of Standards, Gaithersburg, Md. as supplier.
SPP designates Scientific Polymer Products, Inc., Ontario, NY as supplier.
* Supplier values are shown in parenthesis.

Table III. Effect of Sample Concentration on Retention
            Narrow Standards

            Sample: PC-30126   M = 168,000

Conc (% w/v)   Ret.Vol. (ml)

     0.50   13.02
     0.25   12.91
     0.10   12.85
     0.05   12.82
     0.025   12.82

2. Standards from 3 different sources correlate with each other fairly well. Care must be taken not to rely only on the bottle values for standard molecular weights.

3. Intrinsic viscosity determination should be a quick and reliable method for determining the molecular weights of narrow polystyrene standards.

4. Caution must be exercised in making up narrow standards to avoid concentration effects upon retention.

5. The Universal Calibration method is capable of providing accurate values of molecular weight distribution as well as Mark-Houwink parameters from broad distribution samples.

REFERENCES

1. Yau, W.W., Kirkland, J.J., and Bly, D.D., "Modern Size-Exclusion Liquid Chromatography", John Wiley & Sons (1979), p. 252 .

2. Haney, M.A., J.Appl.Poly.Sci., 30, 3023-3036 (1985).

3. Haney, M.A., J.Appl.Poly.Sci., 30, 3037-3049 (1985).

4. Bohdanecky, M., and Kovar, J., " Viscosity of Polymer Solutions", Elsevier (1982), p. 95 .

5. Styring, M.G., Armonas, J.E., and Hamielec, A.E., J.Liq.Chrom., to be published (1987).

RECEIVED April 22, 1987

Chapter 8

Use of a Viscometric Detector for Size Exclusion Chromatography

Characterization of Molecular Weight Distribution and Branching in Polymers

Cheng-Yih Kuo, Theodore Provder, M. E. Koehler, and A. F. Kah

The Glidden Company, Research Center, 16651 Sprague Road, Strongsville, OH 44136

Automated data analysis methodology was developed for an on-line capillary viscometer detector for use with a Waters Model 150C ALC/GPC. Absolute molecular weight distribution curves and statistics were obtained from a universal calibration curve, based on polystyrene, in conjunction with chromatograms from viscometer and refractometer detectors. Yau and Malone's equation was used to fit the non-linear calibration curve. Corrections for dead volume and absolute flow rate also were made. Additionally, Mark-Houwink parameters, K and α, were obtained for polymers of interest and a branching index as a function of molecular weight was calculated for branched polymers. A detailed evaluation of the methodology was performed using various standard and commercial polymers as well as experimental resins. Use of this methodology to generate polymer chain-branching information was demonstrated.

Since its introduction in 1964(1), gel permeation chromatography (GPC) or size exclusion chromatography (SEC) has become the most widely used technique for the determination of the molecular weight distribution (MWD) of polymers. The recent advances in SEC have been centered on the following three areas: a) the development of high performance columns with microparticulate packings for high speed and high resolution separations, b) the interfacing of size exclusion chromatographs to computers for instrument control as well as data acquisition and analysis, c) the development and utilization of molecular size (or molecular weight) sensitive

0097–6156/87/0352–0130$07.00/0
© 1987 American Chemical Society

detectors. There are two types of molecular size sensitive detectors; the low angle laser light scattering photometer and the capillary viscometer. This paper will focus on the capillary viscometer detector.

In the 1970's, quite a number of studies (2-11) were reported involving the coupling of a Ubbelohde type of capillary viscometer to a conventional size exclusion chromatograph (SEC). The polymer fractions eluting from the SEC columns and refractometer were collected in a syphon and then emptied into the Ubbelohde viscometer. The efflux times were used to calculate the intrinsic viscosity of each fraction. This is a discrete batch type operation. With the increased speed and reduced column volumes associated with high performance SEC, this type of viscometer was not practical. In 1972, Ouano (12) first developed a unique on-line viscometer which used a pressure transducer to monitor the pressure drop across a section of a capillary tubing continuously. This continuous viscometer was used with conventional SEC columns. In 1976, Lesec and co-workers (13) described a similar but simpler on-line continuous viscometer detector for their automatic SEC. Evaluation of this viscometer detector for use with a variety of polymer systems was subsequently published.(14) In our laboratory, development of an on-line viscometer detector for SEC was underway concurrent with the work of Lesec and co-workers. The development of our SEC/Viscometer system was first reported on at the 1982 Pittsburgh Conference.(15) In our SEC/Viscometer system a commercially available differential pressure transducer was utilized to monitor the pressure drop across a section of a capillary tubing. To reduce the noise caused by the reciprocating pump systems, two hydraulic dampers were installed. A detailed evaluation of this SEC/Viscometer system was included in a recent ACS SYMPOSIUM SERIES volume.(16) In a recent paper (17) Lesec demonstrated the feasibility of their modified viscometer detector which used two pressure transducers to monitor the pressure drop at each end of the capillary tubing for high temperature work. In 1984, the first commercially available continuous viscosity detector for SEC was introduced by Viscotek.(18) The main component of the Viscotek detector is the Wheatstone bridge configuration consisting of four balanced capillary coils. The detailed description of the design and evaluation of the Viscotek viscometer can be found in a recent publication.(19) Recently Abbott and Yau described the design of a differential pressure transducer capillary viscometer (20) which is comprised of two capillary tubes, one for eluting sample solution and one for eluting solvent. Signals corresponding to each pressure drop measurement are fed to a logarithmic amplifier and the intrinsic viscosity is determined. The measured intrinsic viscosity is independent of flow rate and temperature fluctuations.

In a previous paper (16) the hardware design of the SEC/Viscometer system used in this work has been described. The effects of operational variables, e.g. pump pulsations, flow rate and flow irregularities on the performance of the viscometer and the enhancement of the signal-to-noise ratio by using mechanical dampers also has been described. Our current work has focused on the development, implementation and application of automated data

analysis methodology. In this paper, a detailed evaluation is
given of the computational procedures and methodology used to
characterize molecular weight distribution (MWD) and
chain-branching in a wide variety of polymers.

Instrumentation

Details of the SEC/Viscometer detector system have been described
previously.(16) The key component of the viscometer detector is a
differential pressure transducer (CELESCO Model P-7D, Canoga Park,
CA) with a +25 psi pressure range. The transducer monitors the
pressure drop across a section of stainless steel capillary tubing
(length: 2 ft., I.D. = 0.007 in.). Pump pressure fluctuations
were damped by a method previously described.(16) The viscometer
assembly is placed in the constant temperature column compartment
of the Waters Model 150C ALC/GPC chromatograph between the column
outlet and the refractometer (DRI). The column compartment
temperature was maintained at 40°C. The column set consisted of
six Ultrastyragel columns (2x10^5, 10^4, 10^3, 2x10^6 Å. Waters
Associates, Milford, MA). The mobile phase used was HPLC grade
tetrahydrofuran and the flow rate was set at a nominal flow rate of
1.0 ml/min.
        The absolute flow rate was calibrated using a gravimetric
procedure and a Thermalpulse flow meter (Molytek, Pittsburgh, PA)
which is based on the design of Miller and Small (21, 22). The
operational principle of the flowmeter is based on the measurement
of the time-of-flight of a thermal pulse. The thermal pulse
generated upstream by a thermistor is detected downstream by
another thermistor. The time required for the thermal pulse to
travel through a fixed volume is used to calculate the flow rate as

$$flow\ rate\quad f = \frac{60\ v}{(t-k)} \tag{1}$$

where

    f = flow rate in ml/min.

    v = effective flow cell volume in ml (constant)

    t = time in sec. between pulses

    k = calibration constant (sec).

Each time the downstream thermistor detects the thermal pulse, it
triggers another pulse upstream and the cycle repeats as long as
flow continues. Therefore, it monitors the flow rate continuously.
In our system, the flow rate was determined to be 0.93 ml/min when
the nominal flow rate was set at 1.0 ml/min with THF as mobile
phase operating at 40°C.

Materials

1. The polystyrene standards used for calibration are shown in Table 1 along with the corresponding supplier molecular weight characterization data.

2. The polyvinyl acetate samples (Aldrich 18250-8 Lot #1 and Lot #3) and a polymethyl methacrylate sample (Aldrich 18226-5) were obtained from the Aldrich Chemical Co., Milwaukee, WI.

3. Two polymethyl methacrylate samples (Eastman 6041 and 6036) were obtained from the Eastman Organic Chemicals, Rochester, NY.

4. The polyvinyl chloride (PC-PV-4) sample was obtained from the Pressure Chemical Co., Pittsburgh, PA.

5. Three polydisperse polystyrene samples were used in this study; (i) Dow 1683 was obtained from Dow Chemical Co., Midland, MI, (ii) NB5 706 was obtained from the National Bureau of Standards, Washington, D.C., (iii) PS-4 was a round robin sample from the ASTM Section D-20-70.02 Size Exclusion Chromatography Task Group.

6. The randomly branched polystyrene and two star-shaped polystyrenes were obtained from the Polymer Science Department at the University of Akron, Akron, OH.

TABLE 1. Polystyrene Standards Used for Calibration

Sample *	$\bar{M}n$ x 10^{-3}	$\bar{M}w$ x 10^{-3}	$(\bar{M}n \cdot \bar{M}w)^{1/2}$ x 10^{-3}
WA-27237	2.04	2.12	2.09
PC-11b	3.10	3.60	3.34
PC-8b	9.43	10.0	9.71
PC-41220	15.1	20.5	17.6
PC-7b	33.0	36.0	35.0
ArRo-500-16	97.6	98.1	97.9
NBS-705	170.9	179.3	175.0
PC-1c	193.0	200.0	196.5
ArRo-300-2	392.0	394.0	393.0
PC-13a	640.0	670.0	654.8
PS-16241	945.0	1030.0	987.0
PC-14a	1610.0	1900.0	1749.0
Duke-2575	3727.0	4100.0	3909.0

PC - Pressure Chem. Co., Pittsburgh, PA.
NBS - National Bureau of Standards, Washington, D. C.
ArRo - ArRo Laboratories, Inc., Joliet, IL.
Duke - Duke Standards Co., Palo Alto, CA.
PS - PolySciences, Inc., Warrington, PA.

Data Analysis Methodology

The automation of the SEC/Viscometer detector system is achieved by interfacing the DRI and viscosity detectors to a microcomputer for real time data acquisition. The raw data subsequently are transferred to a minicomputer (DEC PDP-11/44) for storage and data reduction. The automated data acquisition and analysis system for the size exclusion chromatograph with multiple detectors used in this study have been described previously.(23)

Data Reduction Procedures. Details of the data analysis for the SEC/Viscometer system have been described previously.(16) The data reduction scheme is summarized in Figure 1 and briefly will be reviewed here. The intrinsic viscosity [η](V) of the effluent at a given retention volume V is determined from the DRI and continuous viscosity detector responses according to the following equation

$$[\eta](V) = \frac{1}{C(V)} \ell n \left(\frac{\Delta E(V)}{\Delta E_o} \right) \quad C \to 0 \tag{2}$$

where ΔE_o and $\Delta E(V)$ are the viscosity detector responses at constant flow rate corresponding to solvent and to sample having concentration C(V), respectively. For a linear transducer, $\Delta E(V)$ is proportional to the pressure drop across the capillary, $\Delta P(V)$. The concentration C(V) is given by

$$C(V) = W \cdot f(V) / \int_{V_L}^{V_H} f(V) dV \tag{3}$$

Where W (grams) is the weight of the sample injected, and f(V) is the DRI response at the retention volume V. The parameters V_L and V_H represent the lowest and highest values of chromatogram sample retention volume.
The bulk intrinsic viscosity of the sample is given by

$$[\eta] = \int_{V_L}^{V_H} C(V) [\eta](V) dV / \int_{V_L}^{V_H} C(V) dV \tag{4}$$

As shown in Figure 1 data from the viscometer detector and DRI are combined to yield the intrinsic viscosity as a function of retention volume (1a). This curve then was fit to a polynomial and a smoothed curved calculated. At this stage of data reduction the analyst can choose to continue to use the polynomial smoothed values of log [η](V) throughout, or continue to use the unsmoothed values for further data reduction.
    Using the curve log [η](V) vs. V (1a) and the hydrodynamic volume calibration curve, log V_h vs. V (1b), a "secondary" molecular weight calibration curve was generated (1c).

$$M(V) = V_h(V)/[\eta](V) \qquad (5)$$

This "secondary" molecular weight calibration curve was fit to a polynomial over the retention volume range of the sample. Then the molecular weight distribution statistics are calculated from this "secondary" calibration curve and the DRI trace of the sample under analysis.

The Mark-Houwink parameters K and α, where appropriate for the sample of interest, are calculated from a plot of log [η] vs. log M(ld) for linear polymers. For branched polymers, the branching index, g', is calculated from

$$g'(M) = [\eta]_{b,M}/[\eta]_{\ell,M} \qquad (6)$$

where $[\eta]_{b,M}$ is the intrinsic viscosity of the branched polymer at a given molecular weight M, and $[\eta]_{\ell,M}$ is the intrinsic viscosity of the corresponding linear polymer at the same molecular weight M. For polymers with known K and α values, $[\eta]_{\ell,M}$ can be calculated directly.

The above data treatment assumes that the detector cell contents are homogeneous with respect to the molecular size and molecular weight of the polymer species. For highly branched and heterogeneous copolymers this may not be true. Hamielec and coworkers (24,25) have shown that for such cases, a number average molecular weight distribution is derived from the hydrodynamic volume calibration curve. This is particularly important when applying molecular branching models to the data such as that of Zimm and Stockmayer (26) to obtain number and weight-average branching frequency information. In this paper for the polymers studied, branching model calculations were not carried out.

Raw Viscometer Data Smoothing. The computational procedures include Fourier filtering of the raw viscometer data to selectively remove periodic noise at the frequencies of operation of the piston pump system, and high frequency noise from other sources. An example of the Fast Fourier Transform (FFT) smoothing of an exaggerated noisy signal (pump pulse dampers removed) is illustrated in Figure 2 along with the FFT transformed data curve. The program used was the DEC subroutine FFT.(27) In the example shown, 512 discrete data points (corrected for the mean) were passed to the FFT subroutine for a forward transform. The function plotted in Figure 3 shows the square root of the sum of the squares of the real and imaginary portions of the data returned by the FFT subroutine. The abscissa is the index of the point normalized to 2π. Smoothing of the data was effected by setting both the real and imaginary values to zero between 0.4 and (2π-0.4). These end-points were chosen empirically. Since most of the data are described by the first few points, the function appears to be relatively insensitive to the exact choice of truncation end-points. The inverse transform after truncation results in the smoothed curve shown in Figure 2. A typical example is shown in Figure 4 which shows the chromatograms of the polystyrene standard (Mw=17,000) before and after FFT. It is seen that installation of two hydraulic dampers in the pumping system reduces excessive noise

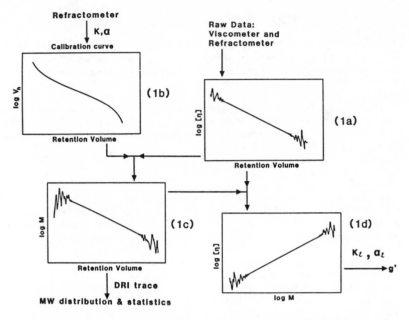

Figure 1. Data Reduction Scheme for Analysis of DRI/Viscometer Chromatograms.

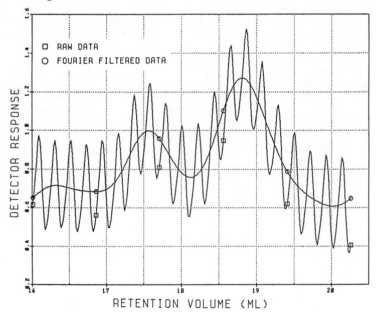

Figure 2. FFT Smoothing of the Viscometer Responses with the Dampers Removed.

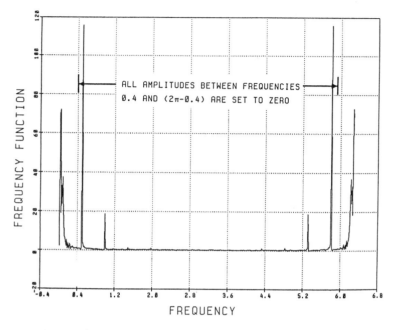

Figure 3. Power Spectrum of the Signal in Figure 2.

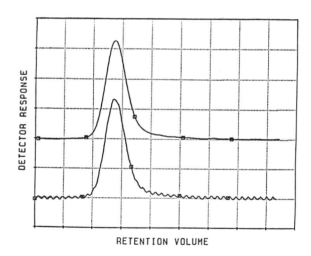

Figure 4. FFT Smoothing of the Viscometer Responses from a Polystyrene Standard.

due to pump pulsations. The FFT analysis of the chromatogram
further smooths and removes the remaining pump pulsation noise;
thereby enhancing the analyst's ability to make baseline
selections.

Non-linear Hydrodynamic Volume Calibration Curve. The hydrodynamic
calibration curve, log V_h vs. V shown in Figure 1b, is generated
using the commercially available narrow MWD polystyrene standards
listed in Table 1 and published values (28, 29) of the Mark-Houwink
parameters K and α for polystyrene in THF at 25°C, (K=1.6 x 10^{-4},
α = 0.706 for M_w > 10,000 and K = 9.0 x 10^{-4}, α = 0.5 for Mw
<10,000). The experimental data points composing the non-linear
calibration curve were fitted with the phenomenologically based
Yau-Malone equation.(30) This equation is derived from diffusion
theory and is expressed as follows:

$$V = A + B \left\{ \frac{1}{\sqrt{\pi}\,\psi} \left[1-\exp\left(-\psi^2\right)\right] + \mathrm{erfc}\,(\psi) \right\} \qquad (7)$$

where $\psi = V_h^D/C$,

and A, B, C and D are constants with A equal to the void volume,
A+B equal to total volume, and erfc(ψ) is the complement error
function of ψ.

The computational procedure for the generation of the
hydrodynamic volume calibration curve requires a set of calibration
values from SEC data (molecular weight, retention volume) which are
fit to the four-parameter Yau-Malone equation using the Nelder-Mead
search procedure.(31-33) The calibration procedure then generates
a curve within the experimental data range, using the values of A,
B, C and D found by the Nelder-Mead procedure. The curve thus
generated is fit to a polynomial to speed up the computational time
during the data analysis. The Yau-Malone equation avoids
inappropriate extrapolations outside the experimental data range
which often occur when a polynomial fit to experimental data points
is extrapolated outside the experimental data range. An example is
illustrated in Figure 5. Other examples demonstrating the use of
the Yau-Malone equation for fitting non-linear calibration curves
can be found in the literature.(34-36)

The effect of polymer concentration on the hydrodynamic volume
also was considered in the generation of the hydrodynamic volume
curve. The computational procedure includes an option for
correcting the concentration effect through Rudin's equation.(37)

$$V_h = \frac{4\pi KM^{1+\alpha}}{9.3 \times 10^{24} + 4\pi\, N_o C\left[KM^{\alpha}-K_\theta M^{0.5}\right]} \qquad (8)$$

where V_h = effective hydrodynamic volume

N_o = Avogadro's number = 6.023×10^{23}
C^o = concentration (g/ml)
K, α = Mark-Houwink parameters
K_θ = K value at theta condition

This equation shows the effect of finite concentration on reducing hydrodynamic volume for very high molecular weight polymers. For polymers with medium molecular weight (ca. 200,000) and below, the concentration effect is minimal. One way to reduce the concentration effect for very high molecular weight polymers is to keep the polymer concentration as low as possible. This also will prevent "viscous fingering" from distorting the chromatograms. For our system it was found that Rudin's correction is not necessary as long as very dilute solutions (ca. 0.01% w/v) were used for the very high molecular weight polystyrene standards.

Dead Volume. The dead volume difference between the viscometer and DRI must be accounted for. Otherwise systematic errors in Mark-Houwink parameters K and α can occur. In the previous paper (16), a method developed by Lesec and co-workers (38) based on injecting a known amount of a very high molecular weight polystyrene standard onto low porosity columns was used. From the viscometer and DRI chromatograms, the apparent intrinsic viscosity [η] was plotted against retention volume V. A series of [η] vs. V plots are then constructed assuming a range of dead volume, ΔV. The slope of each plot is determined by linear regression and is plotted against the assumed ΔV. The correct ΔV corresponds to the [η] vs. V curve having zero slope. The dead volume value, ΔV, was found to be 115$\mu\ell$ for our system using this procedure. The assumption of complete exclusion of the high molecular weight species by the columns might be questionable. A small degree of separation could have taken place in the columns, considering the fact that very high molecular weight polystyrene standards are not strictly monodisperse.

In the current study a different approach was used. The hydrodynamic volume calibration curve was generated with narrow molecular weight distribution polystyrene standards having known Mark-Houwink parmeters. If a broad MWD linear polystyrene sample were analyzed, the measured K and α should agree with the values used to generate the hydrodynamic volume calibration curve providing no significant instrument broadening correction is required. It was found that the measured K and α values were dependent upon the value of the dead volume correction. Figure 6 demonstrates the [η] vs. M dependence on dead volume for a broad molecular weight distribution polystyrene sample. The proper value of dead volume was determined by adjusting ΔV values until the measured K and α values for the broad molecular weight distribution polystyrene sample coincided with the known values for polystyrene. For the chromatographic system discussed in this paper, the best value was found to be 95μl. The Mark-Houwink parameters determined for other linear polymers, using this dead volume value, were in

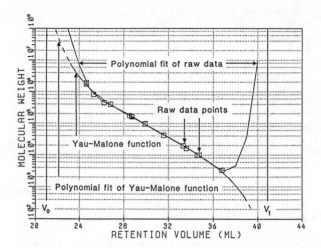

Figure 5. Calibration Curves Fitted with Polynomial and
Yau-Malone Function Followed by Polynomial.

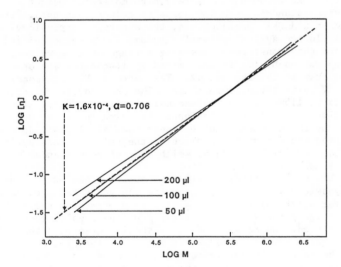

Figure 6. Effect of Dead Volume ΔV on the Slope and Intercept of
the log [η] vs. log M plot.

very good agreement with the literature values. Therefore, this procedure for determining dead volume was found to be self-consistent. Table 2 shows the effect of varying the value of dead volume upon molecular weight averages, intrinsic viscosity, and Mark-Houwink parameters. It is seen that the Mark-Houwink parameters, K and α, are very sensitive to the value of the dead volume between detectors. However, the molecular weight averages and the bulk intrinsic viscosity are barely affected.

TABLE 2. EFFECT OF DEAD VOLUME BETWEEN DETECTORS
(TEST SAMPLE: DOW 1683 PS)

VOLUME ($\mu\ell$)	[η] dl/g	$\bar{M}_w \times 10^{-3}$	$\bar{M}_n \times 10^{-3}$	$K \times 10^4$	α
0	0.86	247	103	0.76	0.765
50	0.86	247	101	1.12	0.730
95	0.86	248	102	1.62	0.704
100	0.86	248	99.3	1.64	0.695
200	0.86	249	101	3.85	0.638

Figure 6 demonstrates that the effect of the dead volume correction is to rotate the [η] vs. M plot about a point [[η],M] which is independent of dead volume. Undercorrecting for dead volume results in values of K which are less than the true value and values of α which are larger than the true value of α. Overcorrecting for dead volume results in values of K larger than the true value of K and values of α less than the true value α.

In this work, using ultrastyragel columns, it was found that instrumental broadening corrections were unnecessary.

Linear Polymers

A series of commercially available polymers have been analyzed with this methodology. Figures 7 and 8 show the DRI and viscometer traces for a PMMA sample (Eastman 6041). Figures 9-11 show the fitted data along with sample data. Although the data at both high and low molecular weight regions appear scattered, the non-noisy data between the vertical bars in Figures 9-11 constitute about 94% of the area under the DRI chromatogram. This is shown in Figure 7 as the crosshatched area. Figure 11 demonstrates the linearity of the log [η] vs. log M curve in the high molecular weight region for a linear polymer. The Mark-Houwink parameters, K and α, are obtained from the intercept and the slope of the straight line, respectively. Figure 12 shows the final molecular weight distribution with cumulative and differential plots. For branched polymers, two additional plots indicative of branching are produced and will be shown later.

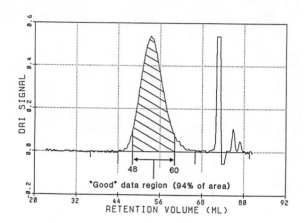

Figure 7. DRI Trace for a PMMA Sample (Eastman 6041).

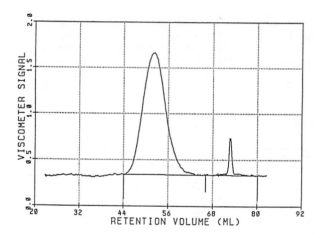

Figure 8. FFT Smoothed Viscometer Trace for a PMMA Sample (Eastman 6041).

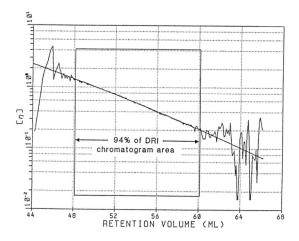

Figure 9. Plot of log [η] vs. Retention Volume for a PMMA Sample (Eastman 6041).

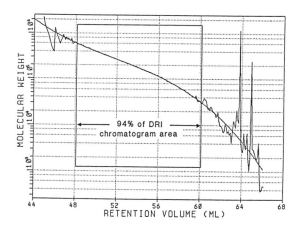

Figure 10. "Secondary" Molecular Weight Calibration Curve for a PMMA Sample (Eastman 6041).

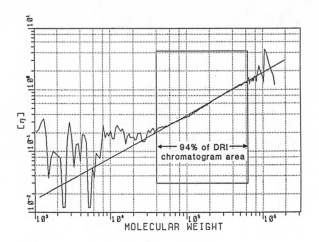

Figure 11. Plot of log [η] vs. log M for a PMMA Sample (Eastman 6041).

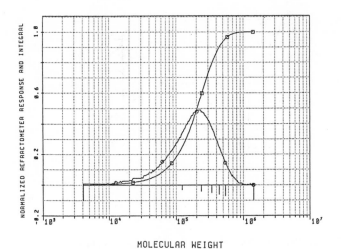

MOLECULAR WEIGHT

Figure 12. Cumulative and Differential Distribution Curves for a PMMA Sample (Eastman 6041).

Polystyrene. Table 3 shows the results obtained for three broad MWD polystyrene samples. The agreement for M_n and M_w values obtained from SEC/Viscometry analysis and the nominal values supplied by the vendors is excellent. In addition the Mark-Houwink parameters, K and α values also are in excellent agreement with each other as well as with literature values.(28,39) This shows the consistency of the analysis method and the technique for determining the dead volume between detectors. The lower M_n value for the NBS 706 sample is due to the low molecular weight tail associated with the sample.

TABLE 3. SEC/VISCOMETRY RESULTS
FOR THREE BROAD MWD POLYSTYRENE SAMPLES

SAMPLE	$\bar{M}_n \times 10^{-3}$	$\bar{M}_w \times 10^{-3}$	$[\eta](dl/g)$	$K \times 10^4$	α
Dow 1683	102	248	0.86	1.62	0.704
	(100)[a]	(250)	–	(1.60)[b]	(0.706)[b]
NBS-706	103	261	0.93	1.47	0.707
	(136)	(258)	(0.93)[c]	(1.60)	(0.706)
ASTM PS-4	103	310	1.06	1.73	0.699
	(105)[d]	(323)[d]	(1.03)[d]	(1.60)	(0.706)

[a] Nominal values supplied by the vendors.
[b] See Reference 28 for literature value.
[c] See Reference 39 for literature value.
[d] ASTM Round Robin results.

Polymethyl Methacrylate. Table 4 shows the results obtained for three commercially available polymethyl methacrylate samples. Again the molecular weight averages obtained with SEC/Viscometry are in good agreement with the nominal values and the K and α values are self-consistent and in excellent agreement with the literature values.(40)

Polyvinyl Chloride. The results obtained for a polyvinyl chloride sample are listed in Table 5. It is seen that the measured molecular weight statistics are in reasonable agreement with the nominal values. The Mark-Houwink parameters K and α obtained from the linear plot of log [η] vs. log M are in good agreement with one group of literature values (41-43) while the α value is lower than that of another group. (3,44-46)

Branched Polymers

Randomly Branched Polystyrene. Branched molecules in solution are more compact than linear molecules and therefore the overall size of a branched polymer molecule in solution is smaller than the

TABLE 4.   SEC/VISCOMETRY RESULTS FOR THREE BROAD MWD PMMA SAMPLES

SAMPLE	$\bar{M}_n \times 10^{-3}$	$\bar{M}_w \times 10^{-3}$	$[\eta](dl/g)$	$K \times 10^4$	α
Eastman 6041	139	242	0.62	1.31	0.686
	(160)[a]	(267)[a]	(0.67)[b]	(1.04)[c]	(0.697)[c]
Eastman 6036	52.6	111	0.374	1.18	0.697
	(48.6)[a]	(115)[a]	(0.371)[b]	(1.04)[c]	(0.697)[c]
Aldrich 18226-5	161	450	1.01	1.24	0.690
	–	–	(1.2)[d]	(1.04)[c]	(0.697)[c]

[a] Nominal values supplied by the vendor.
[b] Measured in Benzene, Polymer Bank Data Sheet.
[c] See Reference 40 for previous literature values.
[d] Inherent viscosity value supplied by vendor.

TABLE 5.   SEC/VISCOMETRY RESULTS FOR A POLYVINYL CHLORIDE SAMPLE

SAMPLE	$\bar{M}_n \times 10^{-3}$	$\bar{M}_w \times 10^{-3}$	$[\eta](dl/g)$	$K \times 10^4$	α
Pressure Chemical PV-4	57	122	1.20	3.36	0.702
	(54)[a]	(132)	(1.25)	(1.63)	(0.766)[b]
				(4.48)	(0.700)[c]

[a] Nominal values supplied by Pressure Chemical Co.
[b] See Reference 44 for literature value.
[c] See Reference 41 for literature value.

linear polymer having the same molecular weight. Figure 13 shows
the plot of log [η] vs. log M for a randomly branched polystyrene
polymer. The deviation from linearity in the high molecular weight
region required a third order polynomial to fit this curve. Figure
14 compares the fitted log [η] vs. log M curve for this branched
polystyrene sample with that of the corresponding linear
polystyrene sample having Mark-Houwink parameters K = 1.6 x 10^{-4}
and α = 0.706. Using Equation 6, the Branching Index, g', can be
calculated as a function of molecular weight and is shown in Figure
15. The molecular weight statistics and the intrinsic viscosity
were calculated to be M_n = 117,000, M_w = 393,000 and [η] = 0.76
dl/g, respectively. These molecular weight averages are close to
those obtained by SEC/LALLS measurement of M_n = 123,000 and M_w =
408,000.(47)

Star-Branched Polystyrene. The SEC/Viscometer methodology was
applied to a narrow 12-arm star-branched polystyrene sample. The
molecular weight of each linear arm is 15,000. The kinetic
molecular weight of this monodisperse star-branched polymer is
180,000. Based on the molecular weight calibration curve generated
from the linear polystyrene standards the molecular weight at the
peak retention volume would be 100,000. Applying the
SEC/Viscometry methodology gave values of M_n = 151,000, M_w =
182,000 and [η] = 0.297 dl/g. Thus, a very accurate M_w value was
obtained from SEC/Viscometry methodology.

The branching index, g', calculated from the intrinsic
viscosity measurement is related to the g value calculated from the
light scattering experiment by the following equations

$$g' = g^\varepsilon \qquad (9)$$

$$g = (\langle s^2 \rangle_{o,b} / \langle s^2 \rangle_{o,\ell})_M \qquad (10)$$

where $\langle s^2 \rangle_{o,b}$ and $\langle s^2 \rangle_{o,\ell}$ are the unperturbed radii of gyration
for a branched and a linear polymer, respectively, of the same
molecular weight. The value of ε lies between 1/2 and 3/2 for a
branched polymer dissolved in good solvent. For a star-shaped
polymer, g can be estimated by using the random walk model (45)

$$g_{R.W.} = (3f-2)/f^2 \qquad (11)$$

where f is the number of branches that radiate from a branch point.
Fetters and co-workers in a recent study (49) on 12- and 18-arm
star polymers found that in theta solvent, g is always greater than
$g_{R.W.}$ suggesting that these star polymers are expanded at the theta
temperature. They also found that in theta solvent, the value of ε
as described in Equation 9 is around 0.62 and seems to be a lower
limit for branched polymers. In good solvents, the values of ε are
higher than that in the theta solvent. This is clearly seen from
Table 6 which lists some of the hydrodynamic data taken from that
paper along with our results. In fact, both values of g' and ε in
THF are in excellent agreement with the values obtained in toluene.

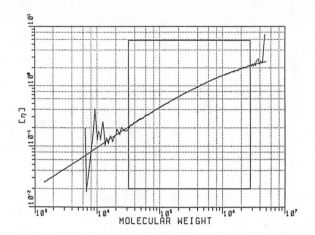

Figure 13. Plot of log [η] vs. log M for a Randomly Branched Polystyrene.

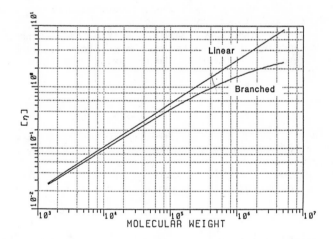

Figure 14. Plots of log [η] vs. log M for a Linear and a Branched Polystyrene.

TABLE 6. HYDRODYNAMIC DATA FOR 12-ARM STAR POLYSTYRENE*

$$g_\theta = \frac{\langle S^2 \rangle_{\theta\,star}}{\langle S^2 \rangle_{\theta\,\ell}} = 0.276$$

$$g_{R.W.} = (3f-2)/f^2 = 0.236 \text{ (Random Walk)}$$

$$g'_{Cyclohexane} = [\eta]star/[\eta]_\ell = 0.41 = g^\varepsilon_{R.W.}; \quad \varepsilon = 0.62$$

$$g'_{Toluene} = [\eta]star/[\eta]_\ell = 0.35 \quad\quad ; \quad \varepsilon = 0.72$$

THIS WORK

$$g'_{THF} = 0.359 \quad\quad ; \quad\quad \varepsilon = 0.71$$

---

* See Reference 49

Figure 16 shows the viscometer and DRI traces of another star-branched polystyrene. This sample contained about 12% of the starting linear arm precursor which eluted at retention volume ca. 52 ml. The kinetic molecular weight of the linear precursor was 260,000. The results obtained for the individual peak through the SEC/Viscosity methodology are summarized in Table 7. It is seen that the measured M_w of the linear arm is very closed to the kinetic value. The average functionality of this star polymer is calculated to be f = 10.

TABLE 7. SEC/VISCOMETRY RESULTS OF STAR PS-W (WHOLE POLYMER)

FRACTION	$\bar{M}_n \times 10^{-3}$	$\bar{M}_w \times 10^{-3}$	$[\eta](dl/g)$	f
Star polymer	1,920	2,530	1.72	10.0
Linear Arm	241	254 (260)[a]	0.914	1.0

$$g' = 0.325 \quad ; \quad \varepsilon = 0.88$$

[a] Kinetic value of molecular weight of linear precursor

Polyvinyl Acetate. Two polyvinyl acetate samples (PVAc #1 and PVAc #3) also were analyzed. Both samples have been shown to be branched by Hamielec (50) by a SEC/LALLS study. Table 8 shows the results obtained from SEC/Viscometry along with some of the available data. It is seen that the intrinsic viscosity values for

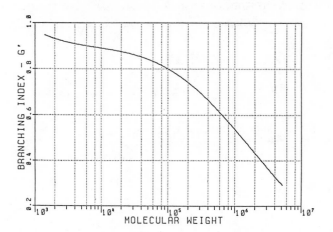

Figure 15. Plot of Branching Index as a Function of Molecular Weight for a Randomly Branched Polystyrene.

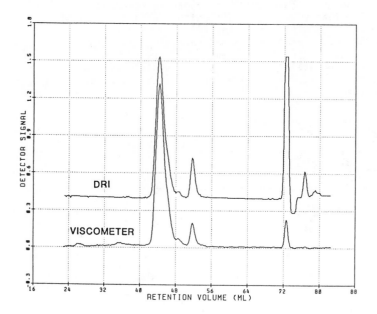

Figure 16. DRI and Viscometer Traces of an Unfractionated Star-shaped Polystyrene.

both samples are in good agreement with the ASTM Round Robin
results obtained with off-line measurements. The molecular weight
averages are comparable with the results obtained from various
sources. The Mark-Houwink parameters, K and α, obtained from
linear extrapolation of log [η] vs. log M, fell between the two
values often cited for linear polyvinyl acetate.(8,51) Dawkins and
co-worker (52) in their most recent publication also found that is
the case. Figure 17 shows the deviation of the plot log [η] vs.
log M from linearity. Also shown in the plot is the molecular
weight distribution curve. The branching index, g', as a function
of molecular weight is shown in Figure 18 along with the molecular
weight distribution curve.

TABLE 8.   SEC/VISCOMETRY RESULTS
OF TWO POLYVINYL ACETATE SAMPLES

PVAC #1

SOURCE	$\bar{M}n$ x 10^{-3}	$\bar{M}w$ x 10^{-3}	[η](dl/g)	K x 10^4	α
SEC/Viscometry(This work)	101	287	0.79	0.89	0.757[a]
Aldrich	83.4	331	-	-	-
Hamielec, et. al.	90.2	300.2	-	-	-
ASTM Round Robin	83.4	263	0.81		

PVAC #3

SOURCE	$\bar{M}n$ x 10^{-3}	$\bar{M}w$ x 10^{-3}	[η](dl/g)	K x 10^4	α
SEC/Viscometry(this work)	109	695	1.48	0.86	0.761[a]
Aldrich	103	840	-	-	-
Hamielec, et. al.	146	626	-	-	-
ASTM Round Robin	102	587	1.51	-	-
Graessley, et. al.[b]				0.51	0.791
Dietz, et. al.[c]				1.56	0.708
Dawkins, et. al.[d]				0.942	0.737

[a] Extrapolated from the linear portion of log [η] vs. log M curve.
[b] See Reference 8.
[c] See Reference 51.
[d] See Reference 52.

Summary

In this paper, the enhancement of the signal-to-noise ratio of the
viscometer signal through the implementation of a numerical FFT
technique is discussed and the computational procedures are
described. A number of examples of quantitative applications to a

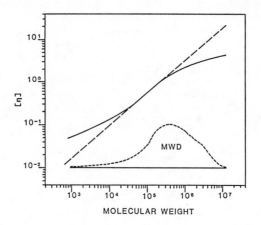

Figure 17. Plot of log [η] vs. log M for a Branched Polyvinyl Acetate (PVAc) (Aldrich 18250-8 Lot #3).

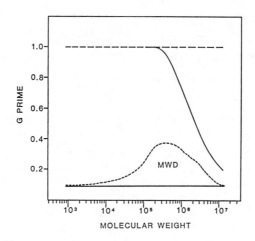

Figure 18. Plot of Branching Index as a Function of Molecular Weight for a Branched PVAc (Aldrich 18250-8 Lot #3).

wide variety of linear and branched polymers are shown. Overall, the SEC/Viscometry detector methodology described in this paper provides very accurate absolute molecular weight distributions, molecular weight averages, bulk intrinsic viscosity values, and Mark-Houwink K and α parameters from a single SEC experiment. In addition, accurate intrinsic viscosity-molecular weight relations can be obtained for linear and branched polymers as well as a branching index as a function of molecular weight.

Literature Cited

1. Moore, J. C., J. Polym. Sci., $\underline{A},\underline{2}$, 835(1964).
2. Meyerhoff, G., Makromol. Chem., $\underline{118}$, 265(1968).
3. Goedhart, D., and Opschoor, A., J. Polym. Sci., $\underline{A2},\underline{8}$, 1227(1970).
4. Meyerhoff, G., Separ. Sci., $\underline{6}$, 239(1971).
5. Grubisic-Gallot, Z., Picot, M., Gramain, P., and Benoit, H., J. Appl. Polym. Sci., $\underline{16}$, 2931(1972).
6. Gallot, Z., Marais, L., and Benoit, H., J. Chromatogr., $\underline{83}$, 363(1973).
7. Servotte, A., and DeBruille, R., Makromol. Chem., $\underline{176}$, 203(1975).
8. Park, W. S., and Grassley, W. W., J. Polym. Sci., Polym. Phys. Ed., $\underline{15}$, 71(1977).
9. Constantin, D., Eur. Polym. J., $\underline{13}$, 907(1977).
10. Janca, J., and Kolinsky, S. J., J. Chromatogr., $\underline{132}$, 187(1977).
11. Janca, J., and Pokorny, S., J. Chromatogr., $\underline{134}$, 273(1977).
12. Ouano, A. C., J. Polym. Sci., $\underline{A-1}$, $\underline{10}$, 2169(1972).
13. Lesec, J., and Quivoron, C., Analusis, $\underline{4}$, 399(1976).
14. Letot, L., Lesec, J., and Quivoron, C., J. Liq. Chromatogr., $\underline{3}$, 427(1980).
15. Malihi, F. B., Koehler, M. E., Kuo, C., and Provder, T., Pittsburgh Conference on Analytical Chemistry and Applied Spectroscopy, Paper No. 806(1982).
16. Malihi, F. B., Kuo, C., Koehler, M. E., Provder, T., and Kah, A. F., "Size Exclusion Chromatography Methodology and Characterization of Polymers and Related Materials", T. Provder, Ed., ACS SYMPOSIUM SERIES NO. 245, 281(1984).
17. Lecacheux, D., Lesec, J., and Quivoron, C., J. Appl. Polym. Sci., $\underline{27}$, 4867(1982).
18. Viscotek Corporation, Porter, Texas.
19. Haney, M. A., J. Appl. Polym Sci., $\underline{30}$, 3037(1985).
20. Abbott, S. D. and Yau, W. W. U. S. Patent 4 578 990, April 1, 1986.
21. Miller, T. E., and Small, H., Anal. Chem., $\underline{54}$, 907(1982).
22. Miller, T. E., Chamberlin, T. A., and Tuinstra, H. E., Am. Lab, January, 1983.
23. Koehler, M. E., Kah, A. F., Niemann, T. F., Kuo, C., and Provder, T., "Size Exclusion Chromatography Methodology and Characterization of Polymers and Related Materials", T. Provder, Ed., ACS SYMPOSIUM SERIES, NO. 245, 57(1984).
24. Hamielec, A. E. and Ouano, A. C., J. Liq. Chromatogr., $\underline{1(1)}$, 111()1978).

25. Foster, G. N., MacRury, T. B., and Hamielec, A. E., in "Liquid Chromatography of Polymers and Related Materials II", J. Cazes, Ed., Dekker, N.Y., 1979, pp.143.
26. Zimm, B. H., and Stockmayer, W. H., J. Chem., Phys., 17, 1301 (1949).
27. Digital Equipment Corporation (Maynard, MA), Laboratory Subroutine Manual #AA-C984A-7.
28. Provder, T. and Rosen, E. M., Separ. Sci., 5(4), 437(1970).
29. Busnel, J. P., Polymer, 23, 137(1982).
30. Yau, W. W., and Malone, C. P., J. Polym. Sci., Polym. Letters Ed., 5, 663(1967).
31. Olsson, D. M., J. Quality Technology, 6, 53(1974).
32. Olsson, D. M., and Nelson, L. S., Technometrics, 17, 45(1975).
33. Smith, J. M., "Mathematical Modeling and Digital Simulation for Engineers and Scientists", John Wiley and Sons, N. Y., 1977.
34. Stickler, M., and Eisenbeiss, F., Eur. Polym. J., 20, 849(1984).
35. Hamielec, A. E., and Meyer, H., Proceedings of ACS Division of Polymeric Materials: Science and Engineering, 51, 541(1984).
36. Rosen, E. M. and Provder, T., Separ. Sci., 5(4), 485(1970).
37. Rudin, A., and Wagner, R. A., J. Appl. Polym. Sci., 20, 1483(1976).
38. Lecacheux, D., and Lesec, J., J. Liq. Chromatogr., 12, 2227 (1982).
39. Hellman, M. Y., in "Liquid Chromatography of Polymers and Related Materials"; J. Cazes, Ed., Dekker, NY, 1977, p. 29.
40. Provder, T., Woodbrey, J. C., and Clark, J. H., Separ. Sci., 6, 101(1971).
41. Samay, G., et al, Makromolecular Chemistry, 72, 185(1978).
42. Takahashi, A., Ohara, M., and Kagawa, I., Kogyu Kagaku Zasshi, 66, 960(1963).
43. Cane, F., and Capaccioli, T., Eur. Polym. J., 14, 185(1978).
44. Freeman, M., and Manning, P. B., J. Polym. Sci., A, 2, 2017(1964).
45. Lyngaae-Jorgensen, J., 7th International GPC Seminar, 1969, p. 188.
46. Bohdanecky, M., Sole, K. Kratochvil, P., Kolinsky, M., Ryska, M., and Lim, D., J. Polym. Sci., A-2, 5, 343(1967).
47. Rudin, A., private communication.
48. Zimm, B. H., and Stockmayer, W. H., J. Chem. Phys., 17, 1301(1949).
49. Roovers, J., Hadjichristidis, N., and Fetters, L. J., Macromolecules, 16, 214(1983).
50. Foster, G. N., MacRury, T. B., and Hamielec, A. E., in "Liquid Chromatography of Polymers and Related Materials II"; J. Cazes and X. Delamare, Eds., Marcell Dekker, Inc., NY, 1980, pp. 143.
51. Atkinson, C. M., and Dietz, R., Eur. Polym. J. 15, 21(1979).
52. Coleman, T. A., and Dawkins, J. V., J. Liq. Chromatogr., 9, 1191(1986).

RECEIVED April 21, 1987

Chapter 9

A New Detector for Determining Polymer Size and Shape in Size Exclusion Chromatography

L. Brower, D. Trowbridge, D. Kim, P. Mukherjee,
R. Seeger, and D. McIntyre

The Institute of Polymer Science, The University of Akron,
Akron, OH 44325

A viscometer has been constructed using membrane pores
as capillary viscometers and can be readily adapted to
conventional SEC equipment. The viscometer has been
shown to give reasonable values of viscosities for
solvent flow and intrinsic viscosities for moderate
molecular weight polymers of MW approximately 100,000.
The viscometer has been used to analyze different
types of polymer structures including microgel. The
microgel pressure difference measurements correlate
with conventional measurements of gel content. The
chromatograms of microgels in both natural rubbers and
commercial acrylic polymers can be obtained and give a
rapid method of detecting microgel. Also the detailed
chromatographic patterns show the possibility of
differentiating between types of microgel. The
observed gel viscometry results are discussed in terms
of polymer entanglements.

Traditional SEC has had difficulties handling extremely large
molecules in which linear polymers have molecular weights exceeding
10^7 g/mol.(1,2). Also the different shapes of large molecules are
difficult to pin point unless there is a marked difference from the
universal calibration.(3) Finally very large molecules commonly
known as microgel (consisting of internally crosslinked and branched
chains of colloidal dimensions) present a scientific challenge to
characterize adequately by any technique. In addition microgel is
also a laboratory hazard to be avoided in a routine SEC measurements
because of its inadvertent plugging of costly columns.

In response to the above characterization problems and an
interest in understanding the topology of intramolecular
entanglement a membrane viscometer was developed.(4) In the membrane
viscometer a solution is passed through a thin (~10 μm) membrane
with well-defined pores of fixed diameter that are nearly
perpendicular to the membrane surface. The important feature is

0097–6156/87/0352–0155$06.00/0
© 1987 American Chemical Society

that the pore (or hole) diameters, D_h, be available in a range of sizes such that the polymer (or molecule) of diameter, D_m, can pass through the large holes freely or the small holes with some difficulty. Both the pressure drop across the membrane and the concentration of effluent are measured continuously while the flow rates of effluent are increased in step-wise or continuous fashion. At this stage of development only the unconfined time-average radius of gyration is considered as ½ D_m. However other viscometric or projection radii may ultimately be more useful.

Early measurements in a steady-state flow apparatus showed that the membrane viscometer allows the direct calculation of kinematic viscosities that are in good agreement with independent capillary viscometer measurements under limited conditions. Agreement is excellent when (1) the average polymer diameter is smaller than the membrane hole, that is, $D_m < D_h$, and (2) the effluent flow rate or its related maximum shear rate, dv/dx, is not too large, $(dv/dx) < 10^3 s^{-1}$.(4) However, it also became clear in the early measurements that to have a generally useful detector for the chromatographic characterization of polymers it would be necessary to avoid concentration polarization and deliver a pulse of polymer through the membrane. To that end an injection loop was used to introduce the polymer solution to the membrane.(5)

The performance of the pulsed flow (injection loop) is similar to that of the unpulsed unit when linear polymer molecules are not too confined by the hole, the shear rates are less than $10^4 s^{-1}$, and concentrations are less than 100 ppm. The pulsed flow apparatus has since been used to explore in a preliminary way the analysis of extremely high molecular weight polymers, entangled polymers, highly branched polymers, and microgels. In this paper a brief description of the apparatus is presented and discussed, and then some interesting preliminary results on microgels are given. Finally a speculative description of the molecular rearrangements that occur during these membrane measurements is followed by a few remarks on the analytical potential of the membrane viscometer detector. Special consideration is given to its use as a general purpose addition to SEC equipment in laboratories already analyzing polymers.

Membrane Viscometer

The membrane viscometer must use a membrane with a sufficiently well-defined pore so that the flow of isolated polymer molecules in solution can be analyzed as Poiseuille flow in a long capillary, whose length/diameter is > 10. As such the viscosity, η, of a Newtonian fluid can be determined by measuring the pressure drop across a single pore of the membrane, knowing in advance: the thickness, L, and cross section, A, of the membrane, the radius of the pore, R_h, the flow rate per pore, Q_i, and the number of pores per unit area, N. The viscosity, the maximum shear stress, σ, and the velocity gradient, $\dot{\gamma}$, can be calculated from laboratory measurements of the above instrumental parameters where $Q_i = Q_{tot}/N$.

$$\eta = \frac{\pi R^4}{8L} \frac{\Delta P}{Q_i}$$

$$\dot{\gamma} = \frac{4}{\pi R^3} \, Q_i$$

$$\sigma = \frac{R}{2L} \, \Delta P$$

Design Concept   The objective of this research was to build an apparatus which could measure the flow behavior of polymer solutions flowing through porous media. The concept of the design was to provide controlled flow to a Nuclepore membrane by using a precision pump and to measure the pressure drop across a single membrane with a sensitive pressure transducer. The dimensions of the flow channels were controlled by selection of the proper pore-size membrane. The concentration of polymer in the solution downstream from the membrane was measured continuously with a differential UV absorbance detector. All solvents passing to the apparatus were prefiltered directly into the primary delivery pump to reduce the possibility of blocking the flow channels of the membrane with dirt and dust. The main advantage of the design was the capability for continuously measuring membrane pressure drop and solution concentration. Thus the apparatus could be used to conduct transient as well as steady state experiments.

Figure 1 presents a diagram of the major sections of the apparatus and their interconnections.

Prefilter Section   The prefilter system for the solvent was a pump and filter holder arrangement similar to that of the measurement system only using a coarse filter of 0.5 μm pores.

Pump Section   Pulseless flow at controlled flow rates was provided by an Isco model 314 pump. This pump is basically a 350 ml motorized syringe with a capability of generating constant flow rates.

Membrane Section   The Nuclepore membranes provided controlled geometry flow channels of polymer molecular dimensions. The pores in a Nuclepore membrane are made by chemically etching the damaged membrane material created when an atomic nucleus passes through a polycarbonate or polyester film. This patented process allows the manufacturer to control the dimensions of the pores simply by controlling the length of time that the film is subjected to etching. The number of pores per unit area of membrane surface is controlled by controlling the flux of nuclei passing through the film during the irradiation process. The pores produced by this process are roughly cylindrical in shape and oriented normal to the plane of the membrane surface(6). The pores are arranged on the surface in a random manner and thus it is possible to have two pores quite close together, however the fraction of pores which exist as doublets or higher multiple pores is less than 10% of the total number of pores. Other workers have measured the pore radii of Nuclepore membranes and found that the reduced standard deviation of the pore radial dimensions is 0.05%.(7)

The Nuclepore membranes used for this work were standard polyester membranes 25 mm in diameter. The pore number, N, is calculated from the pore density and the effective flowing area of

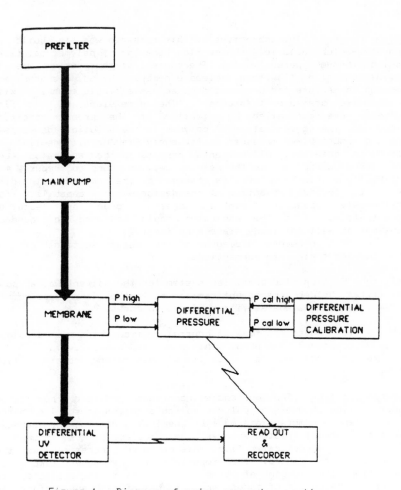

Figure 1.   Diagram of major apparatus sections

the 25 mm membrane in the filter holder which is the area (3.9 cm^2) inside the O-ring seal that holds the membrane. The membranes were mounted in a standard filter holder purchased from Nuclepore. This filter holder was connected to the pump by stainless steel tubing and the filter assembly and inlet tubing were contained in a temperature controlled water bath. Metal tubing and a high pressure pump were used to give the apparatus the capability of running experiments at line pressures above atmospheric pressure downstream of the membrane. However, all of the work reported here was done with the membrane outlet at atmospheric pressure.

Differential Pressure Measurement Section The pressure drop across the membrane is the dependent variable in the experiment and therefore the accurate measurement of small differential pressure is the primary function of the apparatus. The apparatus uses a pressure transducer to make continuous measurements of the pressure drop across the membrane. A Validyne model DP 103 Ultra-Low pressure transducer was selected to make measurements at the lowest pressure drops.

Modified Membrane Viscometer For the pulsed system a coil of tubing (the injection loop) was placed after the prefilter and before the membrane holder as shown in Figure 2. Directional valves at each end of the loop controlled the flow path. Solvent or solution could be pumped directly to the UV to establish baseline absorbance or for calibration. To make P measurements the flow was directed through the membrane and then into the differential UV spectrophotometer. The flow could also be brought to the upstream portion of the membrane holder and then to the UV detector in an effort to measure the concentration at the membrane surface.

Calibration of Instrument When 10 μm and 0.6 μm membranes were used to determine the viscosity of THF using the manufacturer's determination of N from the flow of water, the viscosities of THF were measured to be an average of 85% of the true value. The direct experimental P vs Q curves are shown in Figure 3. (There is, however, a systematic trend below 85% as membranes of even lower pore sizes are used. Although this trend is puzzling it is unimportant for polymer research since most polymer solution studies need relative viscosity, n_{rel}, or specific viscosity, n_{sp}, measurements.)

The n_{rel} data as a function of flow rate, Q, are shown for a 10^5 g/mol molecular weight polystyrene in Figure 4. Both the Ubbelohde viscometric data and the membrane viscometer data are plotted on the same graph for a 0.6 μm pore membrane at a low concentration of 100 ppm. The flow is Newtonian. The actual agreement of the capillary and membrane viscosities at low flow rates is always excellent when $D_m \ll D_h$ and the concentration is extremely low. At small pore size, high concentrations, and high shear rates the flow can become non-Newtonian. The latter effects are only briefly discussed in this paper, but it is this effect that offers an oportunity to characterize the shape rather than the overall size. Even for a relatively large pore (0.6 μm) membrane the shear rates vary from 100 s^{-1} at 2 ml/hr to 10^4 s^{-1} at 200

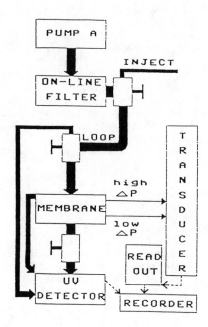

Figure 2.   Diagram of membrane viscometer with loop system

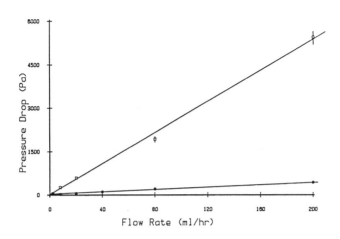

Figure 3. Representative pressure drop vs. flow rate curves for solvent in different membranes: 0.6 μm (●), 0.1 μm (□)

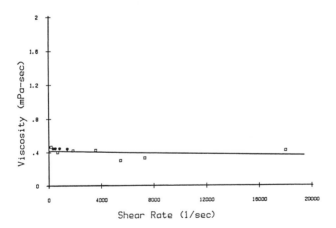

Figure 4. Viscosity vs. shear rate for 10^5 MW polystyrene in membrane viscometer (○) and in glass capillary viscometer (●)

ml/hr, while the pressure differences are still measurable but small
($10-10^3$ Pa), and the maximum shear stresses are extremely low.   The
Reynolds numbers are an order of magnitude below turbulence.   The
D/l ratios are less than 1/10 except for the 10 μm pore membrane   in
which Poiseuille flow formulas need to be corrected for end   effects
to determine true viscosities or shear rates.

Measurements of Gel Content and Type

Conventional Gel Characterization The gels chosen for a preliminary
study were two natural rubber gels   (from guayule, parthenium
argentatum,   (GNR) and hevea braziliensis (HNR) plants)   and   three
commercial acrylic gels of unknown composition from the Glidden Co.
Both types of gel are typical of those encountered in polymer
laboratories and required for product specifications. All   of   the
gels were first characterized both by standard filtration techniques
(using 5 μm and 0.5 μm filters) and by a sedimentation technique,
using a Sorvall centrifuge at 2000 rpm for 0.5 hr. Table 1 shows
the results of the above conventional gel determinations. In each
case the initial concentrations of the solute were 2500   ppm.   In
order to have a standard polyisoprene with little, if any, gel   an
anionically polymerized synthetic polyisoprene rubber (SPI) was
made, tested for gel, and is included in the table. It is apparent
that the natural rubber gel results in Table 1 have a more
meaningful pattern than the acrylic gel results.   This behavior
suggests that the acrylic gels are a different "type" of gel   from
the natural rubber gels. No specific chemical evidence for branch
or crosslink density was determined although such data   must
ultimately be considered in an critical gel analysis. Of course the
microgel is non-uniform in size. Also it is likely that the largest
size are most sensitive to a pore confinement measurement.

Table I. Conventional Gel Determinations

Samples	% Gel		
	Filtration		Sedimentation
	5μm	0.5μm	2000 rpm for ½ hr.
Polyisoprene			
a) anionic	0	0	0
b) guayule rubber (GNR)	2	6	17
c) hevea rubber (HNR)	7	12	--
(SMR-5)			
Acrylics			
A	9	clogged	10
C	0	clogged	25
D	0	37	9
B	(ppt.)		

Membrane Viscometer Characterization of Natural Rubber   The membrane
viscometer was used with a 10 μm membrane in the preliminary
studies.   THF was the solvent and the injection loop contained   4.7
ml.   The chromatograms at a fixed flow rate of 40 ml/hr were   taken
as   a   function of concentration.   At extremely   high   concentration
(1000 ppm) the chromatograms become very broad and take considerable
time   to come back to the baseline.   This is believed to be   due   to
the pair entanglement in concentrated solution and will be discussed
later.   In an effort to determine the point at which this phenomenon
becomes   dominant in a high molecular weight solution the   data   for
the   HNR   are   replotted to be an apparent   viscosity,   ΔP/Q,   as   a
function   of   Q as shown in Figure 5.   Note   that   almost   Newtonian
behavior   is   observed   at   10 ppm.   Maximum   peak   values   of   the
chromatograms   were   used   for the pressure   measurements   shown   in
Figure 6.   The concentration was fixed at 10 ppm in order to be free
of   inter-particle   interactions   from the microgel.   At   such   low
concentrations   there is little excess pressure   difference   (beyond
that due to the polymer solution) developed below a flow rate of   20
ml/hr.   However both the SPI and GNR excess pressures do plateau   by
20 ml/hr, whereas HNR excess pressures continue to grow indefinitely
with increased flow rate.

Membrane   Viscometer Characterization of Acrylic Polymers   The   work
with acrylic gels was extremely complex both in the conventional and
the membrane viscometric characterization as the data in Table 1 and
Figures 7-10 show.   Acrylics appear to have low gel as measured in 5
μm   filters, yet they have a high gel content by sedimentation.   In
addition   a ten-fold decrease in the filter pore size from 5   μm   to
0.05   μm   makes A and C unfilterable.   Also the   gel   contents   from
sedimentation don't   then correlate with   the   filtration   results.
(Another   sample   not shown in the table had a   precipitate   due   to
gravitational   settling   and   therefore   could not   be   run   in   the
membrane viscometer.)

       The   nature of the acrylic gel is obviously quite different from
that   of the natural rubber.   Sample A had the same measureable   gel
in   both   sedimentation and filtration through 5   μm   pore   filters.
Since   pores greater than 10 μm were not on hand, sample A   was   not
run   through   the 10 μm membrane viscometer.   However   the   solution
remaining   above   the   sediment from a sedimentation   of   sample   A,
called A_{sed}, was measured and compared to C.

       The   chromatograms   of   C using a 10   μm   pore   membrane   were
obtained   in the membrane viscometer with 10 μm pores and a   4.7   ml
loop.   The   chromatograms   were   obtained   as   a   function   of
concentration   and   as   a function of flow rate   and   are   shown   in
Figures   7   and 8 respectively.   With these   samples there   is   very
little   pressure built up at low flow rates or   low   concentrations.
This   may   indicate   that the gel is not   a   discretely   crosslinked
colloidal-size molecule but rather a coordinated tangle of molecules
held   together by non-covalent bonds that shift spatially with   time
under   low external stresses.   The assymetry of the chromatograms   at
high   flow   rate   probably   indicates   a   dynamic   response   of   the
entanglement rather than the existence of a distribution of discrete
microgel particles.

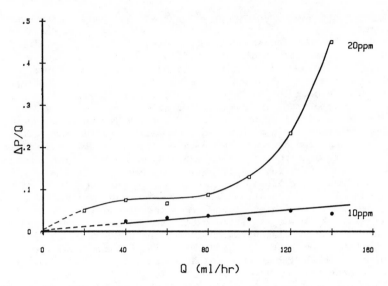

Figure 5. P/Q (arbitrarily scaled) vs. flow rate, Q, for hevea
natural rubber, HNR, solutions containing 8% gel in a total
rubber concentration of 10 ppm (●) and 20 ppm (□)

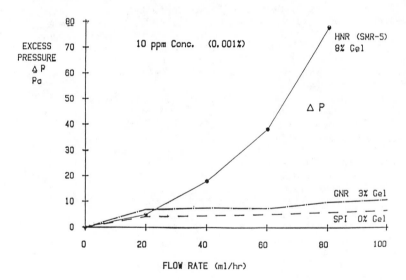

Figure 6.   Excess pressure ,Δp, vs. flow rate for various
polystyrene solutions of hevea, HNR, guayule, GNR, and synthetic
polyisoprene, SPI, with 8%, 3%, and 0% gel determined
gravimetrically and run at a concentration of 10 ppm in THF with
a 10 μm membrane

Concentration Dependency

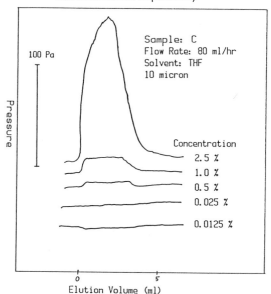

Figure 7. Chromatogram of acrylic gels C as a function of concentration using a flow rate of 80 ml/hr, THF, and a 10 μm membrane

Flow Rate Dependency

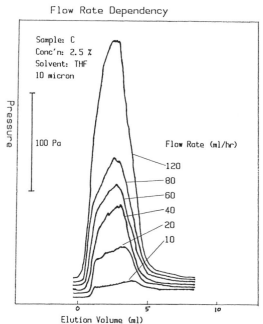

Figure 9. Chromatogram of acrylic gel, C, as a function of flow rate at a concentration of 2.5% in THF using a 10 μm membrane

Flow Rate Dependency

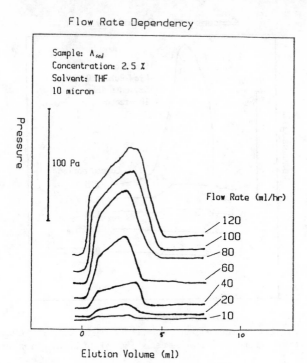

Elution Volume (ml)

Figure 9. Chromatograms of acrylic gel, A_{sed}, as a function of flow rate at 2.5% concentration in THF using a 10 μm membrane

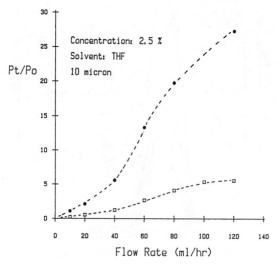

Flow Rate (ml/hr)

Figure 10. Rate of peak heights, p_t/p_o, vs flow rate Q for A_{sed} (□) and C (●) at a 2.5% concentration in THF using a 10 μm membrane

A comparison of A_{sed} and C shows that the A sample, freed of its easily sedimented particles, A_{sed}, has less pressure yet very asymmetric peaks even at low flow rates (Figure 9). Figure 10 with A_{sed} and C compared at the same flow rate and concentration shows that the excess pressure at the peak height for the C sample is always larger than the A_{sed}, and the difference between C and A_{sed} would constantly grow at larger flow rates due to the plateau in A_{sed}. This is best interpreted as the difference in entangling of the entire molecular system even at 2.5% concentration, the concentration at which sedimentation and filtration experiments were made earlier. With this apparatus an estimate of the gel content of the C sample can be made even though it can not be filtered in a 0.5 μm filter. It is not surprising that sample A_{sed} would have less entangling than sample C since the sedimentation had eliminated the most entangled molecular aggregates. It is difficult to interpret the assymetry of the peaks at high flow rates in contrast to the rapid symmetric pressure rise and flat curves at low concentration and low flow rates. It would be best to make more systematic studies of the gels using increasing branching and higher covalent crosslink density as the varying molecular parameters. Until these extensive measurements are made, the role of entanglement intensified by polarization could make the interpretation of results at high concentration difficult. The UV curves correlate with the -curves at low concentrations.

Discussion

The membrane viscometer can rapidly give chromatograms related to the ability of large molecules to reconfigure themselves in order to pass through confining pores. As such the viscometer can be used to elucidate whole chain dynamics involving crosslinks, intra-molecular entanglements, and inter-molecular entanglements. Work is in progress to study these effects systematically in high molecular weight, branched, and model cross-linked systems. However there is also a need to rapidly assess the amount of microgel and the type of microgel in polymeric materials. The present membrane viscometer holds some promise as such a monitoring tool. Also it can readily be adapted to an existing GPC apparatus. In such a configuration a very cheap membrane can save damage to expensive GPC columns and at the same time give an approximation of the amount of gel and an indication of the type of microgel through ΔP measurements as a function of flow rate and concentration.

A greater molecular understanding will be required to interpret the difference between fixed (covalent) and mobile (polar) crosslinks and topological entanglement.

Acknowledgment

This research has been supported in part by the Edison Polymer Inovation (EPIC) Corporation.

Literature Cited

1. E. L. Slagowski, L. J. Fetters, and D. McIntyre, Macromol. 7, 394 (1974).

2.   D. McIntyre, A. L. Shih, J. Savoca, R. Seeger, and A.
      MacArthur, Size Exclusion Chromatography, ACS Symposium 245,
      Edited by T. Provder, ACS, 227, 1984.
3.   M. R. Ambler and D. McIntyre, Polymer Letters 13, 589 (1975).
4.   L. E. Brower, Ph.D. Dissertation, Univ. of Akron, 1985.
5.   D. Trowbridge, M. S. Thesis, Univ of Akron, 1985.
6.   Nuclepore Lab 50 Catalogue, p. 20.
7.   Chauvetau, G. ,J. Rheol. 26, 111 (1982).
8.   M. R. Ambler, J. Appl. Pol. Sci., 20, 2259 (1976).

RECEIVED May 15, 1987

Chapter 10

Determination of Functional Groups in Molecular Components of Polydimethylsiloxanes

Erwin Kohn and Matthew E. Chisum

Development Division, Mason & Hanger—Silas Mason Company, P.O. Box 30020, Amarillo, TX 79177

On-line size exclusion chromatographic (SEC) methods for the determination of silicon hydride, silanol, and silicon phenyl functional groups in molecular weight components of polydimethylsiloxanes of the Sylgard type are described. The methods are illustrated with the analysis results of Sylgards and other commercial samples, some of which have bimodal molecular weight distributions and contain specific functional groups in only one of the molecular weight components. The silicon hydride groups were monitored by means of their IR band near 2160 cm^{-1} and the silanol groups were followed by the hydrogen bonded OH stretching frequency near 3440 cm^{-1}. Silicon phenyl groups were monitored in the UV at 215 nm for quantitation. For silanol and phenyl groups the choice of mobile phase is crucial to the analysis.

Polymeric silicones are extensively used in applications which require thermal stability and long-lasting retention of critical properties. They can be produced in various degrees of hardness and resiliency by combining prepolymer fluids, which contain reactive functional groups, in such ways as to form giant polymer networks with those desired properties.

When designing rubbery materials the aim is to use precursors which are easy to handle, preferably liquid, and which will convert into solids of the proper consistency and other properties, soon after the application is complete. Hardening of the material requires the formation of very large molecules from small ones, which can be accomplished by an addition reaction of functional groups contained in the starting materials. Indiscriminate joining of groups, however, does not result in acceptable properties and careful tailoring of the links is required. This has been accomplished by commercial vendors such as Dow-Corning or General Electric. Nonetheless, production of materials for special applications demanding critical properties requires analytical procedures for evaluation of these commercial precursor fluids. These analytical methods are also needed for an understanding of the curing reaction which is important in the production of such materials.

0097–6156/87/0352–0169$06.00/0
© 1987 American Chemical Society

In this paper detailed methods for the determination of placement
and assay of silicon hydride (Si-H), silicon hydroxide (Si-OH) and
silicon phenyl (Si-∅) functional groups in molecular weight components of silicones of the Sylgard (Dow-Corning Co.) type will be
described. The methods are illustrated with the analysis of Sylgard
addition prepolymers and of model polydimethylsiloxanes (PDMS).
    The Sylgard addition system consists of two fluid silicone pre-
polymers which are mostly PDMSs of relatively low molecular weight.
Part A contains vinyl groups as well as a platinum catalyst and Part
B has both Si-H and vinyl groups. The curing reaction occurs on
mixing of the two liquids, yielding a solid silicone. Physical properties of the silicone depend mainly on the size distribution of
chains in the precursor fluids and on the amount and positioning of
reactive functional groups in the fluids. Other factors, such as
additives and type and concentration of catalyst, are of lesser
importance. The hydrosilation(1) reaction which links the chains
consists of bond formation between the silicon of the hydride group
on one chain, and the carbon of the vinyl group on another chain,
with the simultaneous attachment of the Si-H hydrogen to the vinyl
group (Figure 1). The reaction rate is markedly accelerated by
platinum catalysis and with each such linking event the polymeric
chain or network is greatly increased.
    The position of the functional group on the prepolymer chain is
critical as regards the resulting polymer properties. If the groups
are terminal to short chains or closely spaced on longer chains,
segments between chains will be short, the molecular weight (MW)
will tend to be low and the silicone will have low tensile strength.
If only a single reactive functional group is on some molecules,
nonreactive (terminated) chains will form, resulting in a polymer of
low MW, which is either liquid or a soft solid without strength. If
the functional groups are at the ends of chains (terminal) only,
linear polymers will result which will tend to be soft. Polymers
with more than two reactive groups per chain may form more crosslinked and therefore harder or less resilient silicone products. A
combination of types is frequently desirable to produce materials
with specific degrees of resilience. Positioning of functional
groups is a matter of both the placement of these groups on individual chains and the distribution of chain sizes, or molecular
weight distribution (MWD), of the precursor fluids. A detailed
knowledge of the composition of the precursor fluids with respect to
functional group placement is, therefore, of considerable importance
in establishing critical properties of the silicone products.

Experimental

On-line size exclusion chromatographic (SEC) analyses were performed
with a Waters Model 401 differential refractometer (DR), a Waters
Model 480 ultraviolet (UV) variable wavelength spectrophotometer and
a Foxboro Miran 1A infrared (IR) photometer, equipped with a zinc
selenide ultramicro flowcell of 1.5 mm nominal pathlength and 4.5 μl
volume, purchased from the same supplier. A set of ten Mycrostyra-
gel (Waters Associates) columns, regenerated by Analytical Sciences
Inc. (ASI) and of nominal porosities 100, 500 (two) 10^3 (two), 10^4
(three), 10^5 and 10^6Å, in the order given and a mobile phase flow
rate of 1 ml/min was used. The column set had a specific resolution
of 19.7 in 1,4-dioxane, as determined by the method of Yau(2).

Generally, 1 to 5 mg of sample dissolved in 100 μl of mobile
phase were injected for each SEC run. For the Si-H determinations
tetrachloroethylene (TCE) mobile phase was used at 55°C. It was
Dowper grade (Dow Chemical Co.) and was dried by passage through
activated Molecular Sieves (Linde Co.). The Si-OH analyses employed
1,4-dioxane/ TCE (90/10 by volume) at the same temperature. The
dioxane was purchased from Burdick & Jackson, with UV cutoff of 211
nm and water content of 0.036%. The silicone-phenyl analyses used
methylene chloride (Burdick & Jackson, UV cutoff of 230 nm, water
content 0.003%) at 35°C and tetrahydrofuran (THF, Burdick & Jackson,
UV cutoff 212 nm, water content 0.02% or less) at 50°C. Each of the
mobile phases was continually purged with pure helium during use.
The model silicones (D6170, D6190, T2030, T2078, PS038, PS061,
PS118, PS160, PS340, PS463, PS537, and PS732) were obtained from
Petrarch Systems, Inc. For the IR spectra a Nicolet MX-10 FT-IR
spectrometer was used at a resolution of 1 cm^{-1}. The UV spectra
were measured on a Model DMS 90 Varian Associates UV-Visible spectrophotometer which was connected to an Apple II+ desktop computer,
equipped with an in-house modified Varian Associates software package. Other experimental details are described elsewhere(3).

Molecular Weight Distribution

Size exclusion chromatography was employed to determine the MWD of
Sylgard and other PDMS fluids. The MWD of Sylgard 182B is given by
the bottom line of Figure 2. It shows that the distribution of this
Sylgard prepolymer is bimodal, consisting of a large broad band, representative of a component with average MW of about 1,200 Daltons,
as well as of a smaller band of about 30,000 Daltons. Other Sylgard
prepolymers were also found to have two or more molecular weight
component bands when analyzed in this manner.

Determination of Silicon Hydride Groups

To determine Si-H groups in different molecular size portions of the
sample the Miran IR detector was made part of the SEC analysis system. The instrument was set at 4.64 μm, which is the principal
stretching wavelength of Si-H bonds. The top line in Figure 2 is
the trace from this detector during the SEC analysis of Sylgard
182B. This trace exhibits only a single band, fully corresponding
to the lower MW band detected by the DR. The IR trace corresponding
to the higher MW band has no absorption, which shows that Si-H
groups are absent in this MW component of the prepolymer.
       Quantitative measurements of the lower MW band revealed the
presence of several Si-Hs in each PDMS chain. Furthermore, the
shapes of the DR and IR traces are the same, which indicates that
the distribution of hydride groups is proportional to MW and probably random. An example of a non-random distribution would be a
PDMS with Si-H terminated chains. With two hydrides per chain,
regardless of size, the IR trace of the lower MW portion would be
enhanced and the shapes of the two curves would differ markedly.
This is indeed observed with model PDMSs having hydride groups in
terminal positions. If Si-H positions are random in Sylgard 182B
and there are several Si-Hs per molecule, most of them must be in
internal positions on the chains.

Higher resolution measurements of the Si-H stretching band in model compounds revealed a significant difference in the frequency of this band between terminal and internally placed hydride groups (Figure 3). For terminal hydrides the frequency was about 2127 cm^{-1}, while for internal ones it was about 2167 cm^{-1} (Table I). The frequency for the two Sylgards was about 2163 cm^{-1}, nearly that for internal placements, and in line with a random distribution.

Table I. Silicon Hydride Stretching Frequency for Terminal and Internal Hydrides

Silicone	Si-H Type	Frequency (cm^{-1})
1,1,3,3-Tetramethyldisiloxane (T2030)	Terminal	2128
H-Terminated Polydimethylsiloxane, MW 400 (PS537)	Terminal	2126
Polymethylhydrosiloxane, MW 400 (PS118)	Internal	2167
Sylgard 182B		2162
Sylgard 184B		2163

The size exclusion chromatography-infrared (SEC-IR) method also provided for quantitative determination of Si-H groups in various portions of the MWD of the polymer. This was done by measuring the area under the appropriate portion of the SEC-IR curve which corresponds to a particular MW component. Calibration was performed with a model compound, PS118, which is a PDMS with regularly placed Si-Hs and a MW close to that of the hydride containing Sylgard component. The Si-H content of PS118 was determined by comparing it with 1,1,3,3-tetramethyldisiloxane (T2030) which was found to be of 99+% purity by SEC analysis. Quantitation results for three Sylgard samples for which bulk measurements by nuclear magnetic resonance (NMR) were also available are given in Table II. In the first two samples, 184-31B and 184-36B, the agreement between specific Si-H content by SEC-IR and bulk content by NMR is good. The third sample, 1107-69, shows a lower value by SEC-IR than by NMR. In this case, NMR analysis was performed at an earlier time while SEC-IR was performed much later, after long storage without exclusion of the atmosphere. That a change in the sample had occurred on storage was indicated by its partially gelled state prior to SEC-IR analysis.

Table II. Percent Silicon Hydride Content by SEC-IR and Bulk NMR

Sylgard	SEC-IR	NMR
184-31B	0.45	0.45
184-36B	0.43	0.44
1107-69	1.24	1.60

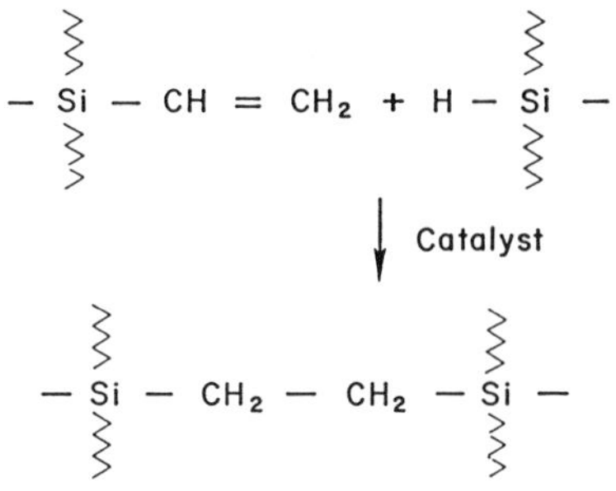

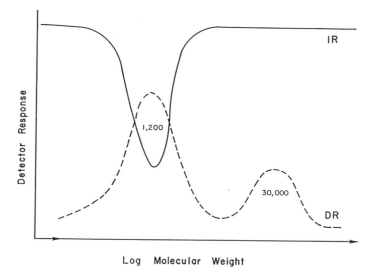

Figure 1. Curing reaction.

Figure 2. DR and IR traces of Sylgard 182B in TCE.

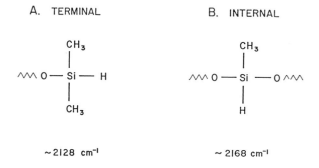

Figure 3. IR of terminal and internal Si-H's in TCE.
A. 1,1,3,3-Tetramethyldisiloxane (T2030). B. Polymethylhydrosiloxane (PS118).

Determination of Silanol Groups

Silanol groups in PDMSs, may form by reaction of Si-Hs with atmospheric moisture or oxygen (Figure 4)(4,5). They interfere with the normal curing reactions through the formation of polymers (Figure 4) of undesirable properties. The free silanol absorbs strongly in the IR at about 3690 cm^{-1}(6) but is partially hydrogen bonded to other OHs(7) and to the oxygens of the polydimethylsiloxane backbone when in a solvent which is incapable of hydrogen bonding. The IR spectrum of the TCE solution of PS340, an OH terminated PDMS with an approximate number average molecular weight of 1,500, is given in Figure 5. The spectrum shows a sharp band near 3690 cm^{-1}, which is assigned to the free hydroxyl absorption, as well as a very broad band centered roughly at 3400 cm^{-1}, which is typical of hydrogen bonded hydroxyls and, therefore, so assigned. The magnitudes of both bands varied with sample concentration and quantitation was not possible.
In a strong proton receiving solvent such as 1,4-dioxane, however, all OH groups are expected to be in the bonded form and a single absorbing species should result. This was indeed found to be the case as shown in Figure 6, which gives the relevant portion of the IR spectrum of PS340 in dioxane. The single broad peak, which is centered near 3440 cm^{-1}, has more than tenfold the area of the TCE absorption and shows no trace of the unbonded species. Silanol concentration in this solvent can be quantitated as illustrated by the calibration curve of Figure 7 in which the absorbance area of the 3440 cm^{-1} peak in PS340 is plotted against microgram of silanol in the mobile phase.
Use of dioxane mobile phase permitted the development of an SEC-IR method for the quantitation of silanol groups in PDMSs. Because the refractive indexes of dioxane and PDMS are nearly the same ten volume percent TCE was added which permitted monitoring of the MWD with the DR detector. Polydimethylsiloxanes not containing hydroxyl groups were found to have a small absorbance at the wavelength of the measurement and correction for this error was made by subtracting the absorbance of PS038, a PDMS of molecular weight comparable to that of the sample and used at the same concentration. The method is illustrated in Figure 8 which shows the SEC-IR and DR traces of a mixture of PS340 and of hexamethyldisiloxane. It is noted that the IR absorption occurs only where the PS340 elutes. Also, the shapes of the IR and DR curves are not the same; the IR curve is more prominent at the lower molecular weights. This is the phenomenon observed before in the silicon hydride analysis; since the silanols are only at both ends of the chains the smaller molecules contain more of them.

Determination of Silicon-Phenyl Groups

Polydimethylsiloxanes containing phenyl groups exhibit absorptions in the UV which are similar to those observed for other phenyl containing compounds(8). They include a principal band in the far UV near 215 nanometer (nm), designated ^1L_a(9) and a weak intensity band multiplet which is centered near 260 nm, designated ^1L_b(9). The latter is often used to recognize and occasionally to quantitate benzenoid content.

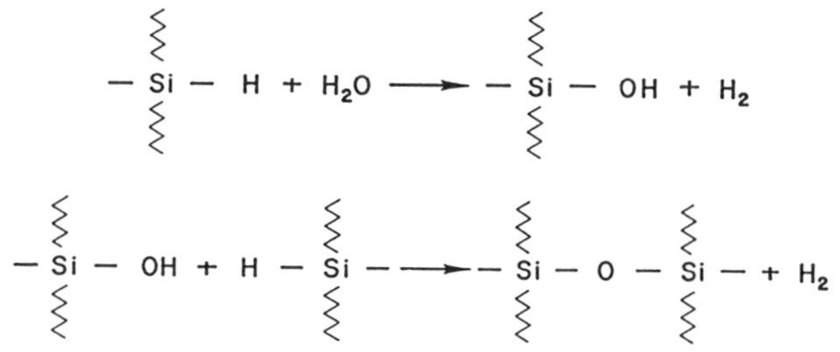

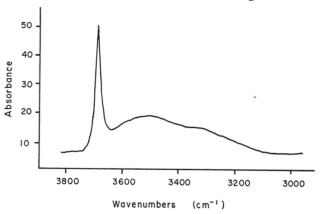

Figure 4.   Reactions of Si-H with H_2O and OH.

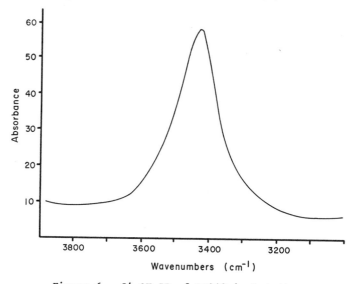

Wavenumbers (cm^{-1})

Figure 5.   Si-OH IR of PS340 in TCE.

Wavenumbers (cm^{-1})

Figure 6.   Si-OH IR of PS340 in 1,4-dioxane.

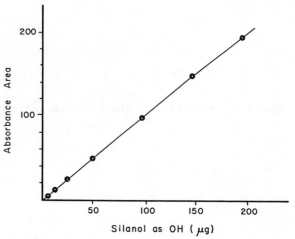

Figure 7. PS340 IR absorbance area of 3440 cm^{-1} band vs microgram of silanol OH in dioxane.

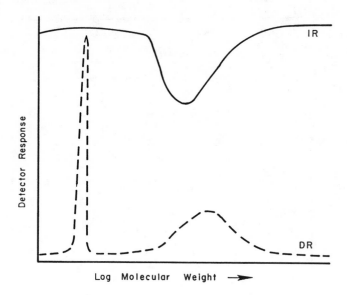

Figure 8.   SEC-DR and IR trace of silicone mixture in 1,4-dioxane.

To qualitatively determine phenyl placements in silicones, SEC-UV analysis was used at 254 nm, with methylene chloride mobile phase. This choice of solvent permits efficient monitoring of the MWD of the sample by employing an IR detector at 9.6 μm, at which wavelength silicones have a strong Si-O-Si asymmetric stretching band. The DR detector is not applicable for this analysis because of similarity in refractive index of methylene chloride and phenyl containing PDMSs. The qualitative aspects of the phenyl analysis are illustrated with Figures 9 and 10, which give the UV (phenyl) and IR (MW) traces of Sylgards 184A and 184B, respectively. It is noted that in each example the higher molecular weight component is relatively rich in phenyl.

Because of the low magnitude of the extinction coefficient of the ^1L_b band the method required relatively large amounts of sample, and may result in "loading" of the columns in case of the low phenyl contents normally observed in Sylgards. Quantitative studies in methylene chloride and other non-polar solvents revealed that the "molar" (formula weight) extinction coefficients of phenyl in these solvents, in both regions of the UV spectrum, vary with the position of the phenyl group in siloxane model compounds and especially if the phenyls are in geminal position(10). The extinction coefficients (E) for the shorter wavelength band in methylene chloride, at or near 230 nm, are given in Table III for a number of phenylmethyl siloxanes.

Table III. Extinction Coefficients for Several Phenylmethyl
Siloxanes at or Near 230 nm in Methylene Chloride

Siloxane	E(*)
1,3-Diphenyltetramethyldisiloxane (D6190)	250
1,1,3,3-Tetraphenyldimethylsiloxane (T2078)	600
1,3-Diphenyltetrakis-1,1,3,3-(dimethylsiloxy) disiloxane (D6170)	50
Polydimethylsiloxane, vinylphenyl terminated (PS463)	280
Copolymer-(96%)dimethyl-(4%)diphenylsiloxane (PS732)	730
Polymethylphenylsiloxane (PS160)	150

(*)Per formula weight of phenyl

Since the intensities of the bands differ for different placements of the phenyl group in the silicones, the method does not lend itself to quantitation. The intensity of the shorter wavelength band, however, is much enhanced in polar solvents, such as 1,4-dioxane and THF, and is less variable with changes in position of the phenyl groups along the PDMS backbone. This intensity enhancement is probably due to charge-transfer complexing of phenyls with the oxygens of the solvent. In these complexes the electronic transitions are less influenced by the position of the phenyl group along the siloxane chain than they are in non-polar solvents where complexing is weak or absent. The extinction coefficients for the shorter wavelength band in dioxane and THF, at or near 215 nm, are

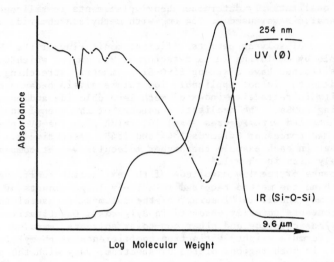

Figure 9.   SEC-IR and UV trace of Sylgard 184A.   IR at 9.6 m, UV
at 254 nm.

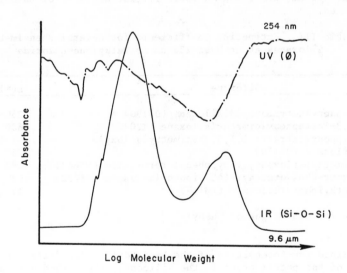

Figure 10.   SEC-IR and UV trace of Sylgard 184B.   IR at 9.6 m, UV
at 254 nm.

shown in Table IV for a number of phenylmethyl siloxanes. More than
a ten-fold increase in E for dioxane or THF versus methylene
chloride was observed for most siloxanes. The UV spectra of
1,3-diphenyltetramethyldisiloxane and of Sylgard 184A in THF are
shown in Figures 11 and 12.

Table IV. Extinction Coefficients for Several Phenylmethyl
Siloxanes at or Near 215 nm in 1,4-Dioxane or THF

	E(*)	
Siloxane	Dioxane	THF
1,3-Diphenyltetramethyldisiloxane (D6190)	9,200	9,200
1,1,3,3-Tetraphenyl-1,3-Dimethyldisiloxane (T2078)	8,400	8,200
1,3-Diphenyltetrakis-1,1,3,3-(dimethylsiloxy) disiloxane(D6170)	6,400	5,800
Polydimethylsiloxane, vinylphenyl terminated (PS463)	8,000	8,700
Copolymer-(96%)dimethyl-(4%)diphenylsiloxane (PS732)	6,100	–
Polymethylphenylsiloxane (PS160)	7,900	9,400

(*) Per formula weight of phenyl

    Tetrahydrofuran was chosen for the SEC analyses because its re-
fractive index provides a slightly larger difference for detection
of the samples and monitoring of the MWD, than does dioxane. But
the difference in refractive index is still not fully satisfactory
for good detection of the MW of the dimethylsiloxane backbone.
Representative SEC-UV and DR curves for Sylgard 184A in THF mobile
phase are given in Figure 13. A comparison of the shapes of the
curves representing phenyl content as detected by UV absorbance at
215 nm and the MWD as determined by the DR, shows a greater concentration of phenyls at the lower MW end of the distribution. If an
unusual distribution of phenyls is excluded the results suggest
terminal placement of phenyls in these Sylgards, as was observed in
certain silanol containing compounds described previously. PS463, a
PDMS with terminally placed phenyl groups, shows this effect, while
another, PS061, which has a small internal block of phenylmethyl-
siloxane units, does not.
    Quantitation of phenyl content in Sylgards was performed by
comparison of areas of UV absorbance at 215 nm with a calibration
curve generated with PS463, a phenyl terminated PDMS. In performing
the integration for phenyl content the absorbance due to the THF
peroxide band was omitted but other bands, due to lower molecular
weight impurities which contain phenyls, were included to permit
comparisons with bulk analytical methods, such as NMR.
    Interference by functional groups other than Si-H, which are
present in many Sylgard prepolymers, has not been investigated. It
is expected that aromatic groups such as biphenyl and naphthyl will
give absorbances similar to phenyl. The absorbance of Si-H containing PDMSs at 215 nm was negligible.

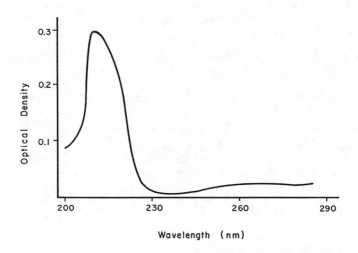

Figure 11.   UV spectrum of 1,3-diphenyltetramethyldisiloxane
(D6190) in THF.

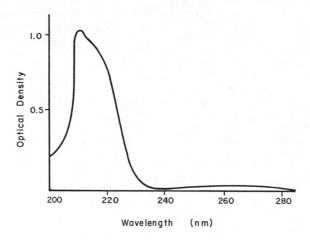

Figure 12.   UV spectrum of Sylgard 184A in THF.

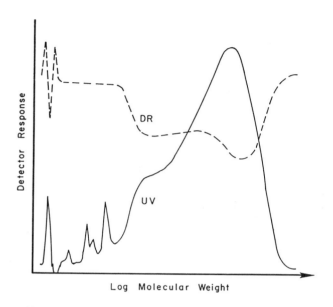

Figure 13.   SEC-UV at 215 nm and DR of Sylgard 184A in THF.

Conclusions

Methods for the detection of placement and quantitation of Si-H,
Si-OH and Si-∅ groups in molecular weight components of silicones of
the polydimethylsiloxane type have been developed using SEC-IR and
SEC-UV on-line techniques. In the case of silanols, problems with
partial hydrogen bonding were overcome by use of 1,4-dioxane mobile
phase. For phenyl groups, methylene chloride was effective for
qualitative measurements and THF yielded quantitative results at
high sensitivity.

Acknowledgments

We are indebted to George L. Clink for the NMR analyses and to W.T.
Quinlin and R. S. Skelton for help in adapting the UV spectrometer
software. The work was performed under auspices of the Department
of Energy, Contract No. DE-AC04-76DP-00487.

Literature Cited

1.   Eaborn, C.; "Organosilicon Compounds"; Academic: New York,
     1960; pp. 45-64.
2.   Yau, W. W.; Kirkland, J. J.; Bly, D. D. "Modern Size Exclusion
     Liquid Chromatography"; Wiley: New York, 1977.
3.   Kohn, E.; "High Performance Size Exclusion Chromatography (IX)
     Bulk and Component Analysis of Silanol Groups in Polydimethyl-
     siloxanes"; Mason & Hanger - Silas Mason Co. Inc., Amarillo,
     TX, MHSMP-85-06 (January 1985). Kohn, E.; "High Performance
     Size Exclusion Chromatography (VII) Determination of Silicon
     Hydride Functional Groups in Components of Silicones"; ibid,
     MHSMP-83-28 (August 1983). Both available through National
     Technical Information Service (NTIS).
4.   Noll, W.; "Chemistry and Technology of Silicones"; Academic:
     New York, 1968, p. 90.
5.   Voorhoeve, R. J. H. "Organosilanes Precursors to Silicones";
     Elsevier: 1967, p. 295.
6.   Smith, A. L. "Analysis of Silicones"; John Wiley; New York,
     1974; p. 275.
7.   Voronkov, M. G.; Mileshkevich, V. P.; Yuzhelevskii, Yu. A. "The
     Siloxane Bond"; Consultants Bureau: New York, 1978, p. 69.
8.   Ramsey, B. G.; "Electronic Transitions in Organometalloids";
     Academic: New York, 1969.
9.   Jaffe, H. H.; Orchin, M. "Theory and Applications of Ultra-
     violet Spectroscopy"; Wiley: New York, 1966.
10. Uriu, T.; Hakamada, T. Kogio Kaqaki Zasshi 1959, 62, 1421.

RECEIVED August 18, 1987

Chapter 11

Analysis of Coal Liquids by Size Exclusion Chromatography–Gas Chromatography–Mass Spectrometry

C. V. Philip, P. K. Moore, and R. G. Anthony

Kinetic, Catalysis, and Reaction Engineering Laboratory, Department of Chemical Engineering, Texas A&M University, College Station, TX 77843

Size exclusion chromatography (SEC) using a column similar to a 5 micron, 100A PL-Gel column (7.8 mm id x 60 cm long) and dry, additive-free tetrahydrofuran (THF) as the mobile phase, separates coal liquids into fractions containing species with similar linear molecular sizes. Since most similar molecular size species in coal liquid happen to have similar functionalities, the size separation enables the separation of fractions containing similar chemical species. The fractions separated by SEC are collected by a multiloop sample valve. The multiloop valve links the liquid chromatograph with a high resolution gas chromatograph. The GC system is equipped with a Finnigan Ion Trap Detector (ITD), a mass spectrometer for capillary chromatography. The volume of the fractions as well as the timing of fraction collection can be varied in order to study any specific species in the sample. The technique can be used to analyze coal liquids, petroleum crudes, and their various distillation cuts.
   The analysis of Wyodak recycle solvent by SEC-GC-MS shows that certain distribution order exists for species in coal liquids with respect to their size and degree of isomerization. The alkanes increase in chain lengths without any appreciable degree of isomerization, except for some biological markers such as pristane and phytane. Phenols and aromatics vary in size and extent of isomerization which causes the liquid to contain a large number of species.

0097–6156/87/0352–0183$06.00/0
© 1987 American Chemical Society

An enormous amount of work both at bench scale and at pilot plant scale have been conducted to study the production of liquid and gaseous hydrocarbons from coal. Since most of the analytical methods are either very time consuming or very specialized, almost all the data available on the coal liquefaction process are based on distillation data or on the assumption that all products which are not insoluble solids are converted. It is known that products of liquefaction vary based on coal, reaction conditions, and media of reaction; hence, conversion and yield may be based on very different products.

The type of quantitative analytical data which are needed for modelling and kinetic studies on coal liquefaction process could not be obtained by using general analytical techniques. We have developed a new analytical approach for obtaining qualitative information as well as quantitative data on coal liquid species. Coal liquefaction produces smaller molecules from coal which is composed of larger molecular species or a matrix of larger molecular species in which smaller species are entrapped.

Size exclusion chromatography (SEC) separates molecules based on size in a short analysis time. Unlike other chromatographic techniques, SEC does not retain sample species in the column, the analysis time is fixed, and everything loaded onto the column elutes within a fixed time frame. The application of SEC is limited only to the solubility of the sample in a solvent. Since tetrahydrofuran (THF) is a good solvent for coal liquids, the separation of coal liquids by SEC can be easily achieved.

Although the well established use of size exclusion chromatography (SEC) is the separation and characterization of polymers based on molecular size or molecular weight, its use can be extended for the separation of smaller size molecules (1-4). Since coal derived mixtures have several small molecules of similar sizes, the use of SEC alone does not resolve them for the purpose of identification. The gel columns packed with 5μm polymer particles have about 50,000 theoretical plates per meter, a five fold increase over that of 10 μm columns; thus, an increase in separation efficiency is achieved. Because sample sizes can be increased with reasonably good resolution, and a 60 cm long column can separate a relatively large sample in a time as short as 25 minutes, SEC can be used as a preliminary separation technique prior to use of other analytical techniques such as gas chromatography (GC) and gas chromatography-mass spectroscopy (GC-MS).

Coal liquids, petroleum crudes, and their distillation cuts have been separated into four or five fractions by SEC (5-15). The SEC fractions were analyzed by use of GC. The procedure was performed manually. It was inefficient, and susceptible to human error. The automated fraction collection followed by injection of the fraction into the GC reduces analysis time, and offers an option for collection of the desired number of fractions at predetermined time intervals. The manual collection of up to 10 one-ml fractions is also used in order to study the effectiveness of the automated method.

The flame ionization detector (FID) can be used for the detection and quantitative estimation of components separated by the GC. Identification of major species can be achieved by a mass spectrometer which can not be used for quantitative analysis of complex mixtures such as coal liquids.
Mass spectrometers used to be expensive and complex for routine use as a GC detector. The Ion Trap Detector (ITD, Finnigan) is a low priced mass spectrometer (MS) for capillary chromatography. Three analytical tools - SEC, GC, and ITD - are incorporated into a powerful analytical system for the analysis of complex mixtures such as coal liquids, petroleum crude and various refinery products. The instrumentation and the SEC-GC-MS analysis of a coal liquid are presented in this paper in order to demonstrate the technology.

EXPERIMENTAL

Coal Liquid Samples

Coal liquids used in this work are recycle solvents which were obtained from the University of North Dakota Energy Research Center where anthracene oil was used to liquefy coals and the resulting product slurries   were recycled as liquefaction solvents approximately twenty times. The recycle solvent contained a substantial amount of original anthracene oil and its decomposition products along with coal derived products. The recycle solvent represents a very complex synthetic crude sample. The recycle solvents produced from Wyodak subbituminous coal, two North Dakota lignites (ZAP-2 and Beulah), and Texas Big Brown lignite were used for the analysis. These recycle solvents were used for mini-reactor liquefaction experiments at Texas A&M University. The products from these experiments are also being analyzed. A 25% solution of the crude was prepared in tetrahydrofuran (THF) and filtered through 0.45 μm disposable HPLC filters (Supelco) and 100 μls of the filtered solution were used for each SEC separation.

SEC-GC-MS Instrumentation

A schematic of the system is shown in Figure 1. The system consists of the following: a liquid chromatograph (LC$_{\text{b}}$, Waters ALC/SEC Model 202) equipped with a 60 cm, 5 μm, 100 Å PL-gel column (Polymer Laboratories) and a refractive index detector (Waters Model R 401); a gas chromatograph (Varian, VISTA 44) equipped with a bonded phase fused silica capillary column (BP5, 0.32mm ID, 25M long) manufactured by Scientific Glass Engineering, Inc. (SGE); an autosampler (Varian 8000); a flame ionization detector (FID); a nitrogen specific detector (Thermionic Ion Specific - TSD); and a microcomputer system (IBM CS 9000) with 1000K bytes RAM and dual 8" floppy disk drives for collecting raw chromatographic data.
A Finnigan Ion Trap Detector (ITD), a small mass spectrometer for capillary chromatography, is the third detector interfaced with the gas chromatograph. The control of the ITD, the data collection, and the identification of species, by a library

search with a NBS Mass Spectral Data Base stored on a 10 megabyte hard disk, are performed by an IBM PC XT microcomputer. Since the currently available software does not allow the operation of the ITD in a run programmed mode, it is manually reset between GC runs. New software for ITD will be available soon for uninterrupted analysis of several samples. Frequently samples collected on the loops of valve V_3 (Figure 2) were concentrated and manually injected as a larger-size sample (0.2-0.5 µl) with a Scientific Glass Engineering (SGE) on-column injector in order to obtain a good mass spectral fragmentation pattern of minor components. After the initial identification, the major fragmentation peak is used in routine SEC-Ge-MS analysis.

SEC-GC Interface

The continous sample separations on the gel column followed by the GC analysis of selected fractions were achieved by the operation of two 6-port valves and a 34-port valve (All from Valco Instrument Company) as illustrated in Figure 2. Sample injection into the LC was performed by a sixport valve (V_1) with a 2 ml sample loop and fitted with a syringe-needleport for variable sample size injection. The combined operation of another 6-port switching valve and the 34-port valve (V_3) with 16 sample loops (100 µl) enabled the linking of the liquid chromatograph with the autosampler of the gas chromatograph. The autosampler was modified to handle 100 µl samples directly from the fraction collection loops of V_3. When V_2 was turned clockwise, it kept V_3 in line of the LC effluent so that the fractions of separated sample could be collected and the autosampler was bypassed. V_2 at its counter clockwise position kept V_3 in line with the autosampler for sample injection but it bypassed the LC stream. Generally, 0.1 µl samples were used for the GC analysis. Sometimes the stream from the capillary column was split (50/50) for the simultaneous monitoring by the FID and TSD. The real time monitoring of the GC was possible on both Varian and IBM systems and the raw chromatographic data were stored on the 8" floppy disks. The fraction collections and sample injections into the gas chromatograph, as well as the data collection, were performed by the integrated system composed of a Varian Automation System (VISTA 401) and the IBM microcomputer (CS 9000). For each sample injected into the SEC column, up to 16 fractions were collected and analyzed by the GC using appropriate gas chromatographic programs stored in the memory without any manual interaction.

In addition to the use of the LC-GC interface (Figure 2), 1 ml fractions were manually collected and analyzed by gas chromatograph using the autosampler. The concentrated fractions were also evaporated using a slow stream of nitrogen and analyzed each fraction by GC-MS using a SGE on-column injector.

RESULTS AND DISCUSSION

Wyodak Recycle Solvent

Figure 3 shows the SEC separation of Wyodak recycle solvent. The

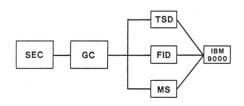

Figure 1.   SEC-GC-MS instrumentation.

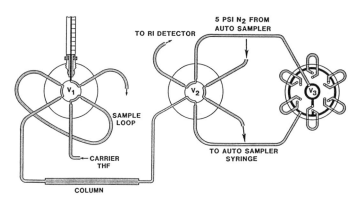

Figure 2.   SEC-GC interface Note: V_3 has sixteen sample loops instead of six shown.

separation pattern of various chemical species and chemical
groups are assigned based on reported (4-15) as well as
unreported studies. When valves V_2 and V_3 (Figure 2) are
engaged, the SEC effluents are collected in the sample loops of
V_3 at specific intervals. The refractive index detector output
shows the effect of such fraction collections as negative peaks
(Figure 4).

SEC Separation of Major Chemical Species in Coal Liquids

The major species in a coal liquid can be "lumped" as nonvolatiles, alkanes, phenolics, aromatics and other species, including
unidentified species. The order in which they elute from SEC is
illustrated in Figure 3. Within each group the species are
separated in the order of decreasing linear molecular size. The
"nonvolatives" include heavy molecules and the species that
accumulate on the capillary GC column and are thus not detected
by FID and ITD. The alkanes are mostly straight chain
hydrocarbons ranging from $n-C_{14}H_{30}$ to $n-C_{44}H_{80}$. Alkanes larger
than $n-C_{30}H_{62}$ were not usually detected due to the conditions
used in most of this work. Their presence was established by
conducting one or two analysis by expanding the limits of the GC
temperature program. The next chemical lump that elutes from
SEC are phenolics, which include all alkylated aromatic compounds
with the hydroxy functional-group, such as alkylated phenols,
indanols, and naphthols. They elute from the SEC in the order of
decreasing linear molecular sizes. The last major chemical lump
to elute from the SEC are the aromatics, and they include
alkylated-benzenes, indans, naphthalenes, and heavier polycyclic
aromatics such as pyrenes, dibenzofuran, and dibenzothiophenes.

Gaschromatograph Analysis of SEC Fractions of Wyodak Recycle
Solvent

Sixteen SEC fractions of 100 μl each were collected from the
Wyodak recycle solvent (Figure 4) at 0.5 min intervals. Each
fraction was analyzed by injecting 0.1 μl into the GC using the
flame ionization detector (FID). The first three and the last
samples did not show any peaks other than the peaks derived from
the solvent; so the GC of those fractions are not included in
Figure 5. The first GC (Figure 5-1) corresponds to the GC of
fraction #4 and the last GC (Figure 5-12) is that of fraction
#15. By increasing the GC oven temperature the larger alkanes in
fraction #2 and #3 can be detected. A shorter column enhances
the FID response as these heavy alkanes accumulate on the column
probably due to irreversible adsorption or decomposition.

Identification of Species in SEC Fractions of Wyodak Recycle
Solvent

Figure 5-1 shows the GC of fraction #4. It shows alkanes ranging
from C_{25} to C_{30}. It is quite possible that the fraction may
contain higher alkanes which are not detected due to the GC-oven
temperature limit. The peaks are identified from the MS fragmentation pattern. Fraction #5 is collected after a 0.5 minute

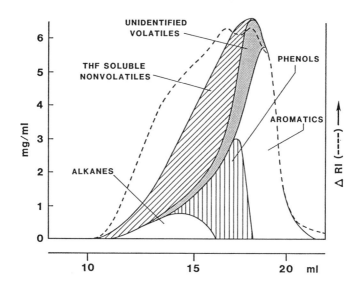

Figure 3.   SEC of Wyodak recycle solvent

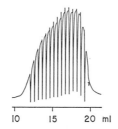

Figure 4.   100 μl Frs. Collected by SEC-GC interface (on-line)

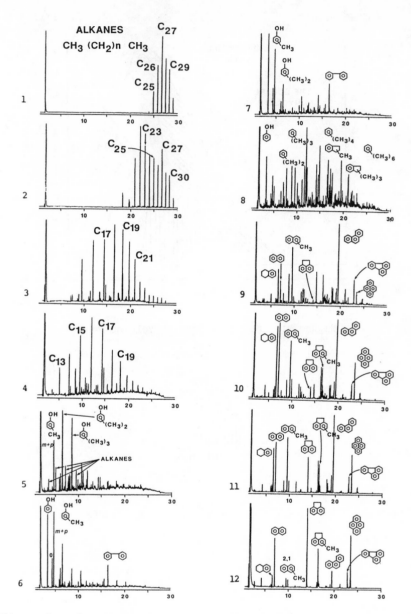

Figure 5. Gaschromatograms of SEC fractions (Fr. #3 through #14 in Figure 4). Temperature program 80°–120°C at 4°C/min, 120–280°C at 10°C/min and 4 min. hold at 280°C.

interval and its GC (Figure 5-2) shows alkanes ranging from C_{19} to C_{30}. This fraction has smaller alkanes ($C_{19}-C_{24}$) in larger proportions in addition to smaller amounts of alkanes ($C_{25}-C_{30}$) which were detected earlier in fraction #4.

When SEC columns are overloaded the peaks tail to longer retention time or volume. In the case of other modes of chromatography, such as gas chromatography, the overloading causes a shift in the retention time towards lower values, and peaks skew towards lower retention times. A "cascading effect" occurs due to the decreasing number of active sites seen by the sample as the number of the same sample molecules, which temporarily block the active sites, increases. The reverse phenomenon is true in the case of SEC where a larger sample size causes peak spreading and the sample to elute over a slightly longer retention time period.

The peak due to C_{27} is larger relatively to other alkanes in Figures 5-1 and 5-2. Our experience with fossil fuels indicates that the straight chain alkanes (n-alkanes) have a normal distribution over a wide molecular weight range. Even in other reported works on hydrocarbon fuels an unusual enhancement of a particular straight chain alkane is not observed. The alkane fractions always contain branched alkanes such as pristane, phytane, and hopane, and some of them are called biological markers. It is quite possible that the C_{27} peak could be due to some branched alkanes co-eluting with n-C_{27}.

The GC of fraction #6 is shown in Figure 5-3, which contains mostly alkanes in the range of $C_{15}-C_{24}$ and small amounts of C_{14} and $C_{25}-C_{29}$. The fraction #6 was collected 0.5 minutes after fraction #5 and 1 minute after fraction #4. If fraction #5 were not collected, the GC of fraction #4 and #6 which were 1 minute apart would have identified all the species but not necessarily quantitatively. Fraction #5 has species from both fractions #4 and #6. The peak width of species eluting at these retention times is about 1 minute. Hence SEC fraction collection at 1 min (1 ml) intervals would have identified all the species with less overlapping and analysis time could have been reduced to half. A short peak immediately after C_{17} is pristane (trimethylhexadecane). The short peaks that appear between the n-alkane peaks appear to be isoalkanes or branched alkanes. The base line appears to be shifted slightly upward compared to the base line of the GC's in Figure 5-1 and 5-2. This is probably due to a large number of possible isomers of phenolic species.

GC of fraction #7 (Figure 5-4) has alkanes as small as C_{12}. The ratio of peak heights of pristane to C_{17} increases in the GC of this fraction compared to previous fractions as expected from its shorter linear molecular size. The smaller peaks between n-alkane peaks are alkylated phenols and branched alkanes.

Fraction #8 (Figure 5-5) is mostly alkylated phenols and indanols with a trace amount of smaller alkanes. The base line shift is due to the co-elution of several large phenolic species in many isomeric forms. Fraction #9 (Figure 5-6) does not contain any alkanes. The ratio of the o-cresols to m, p-cresols increases from fraction #8 to #9. Both m-cresol and p-cresol are structurally longer than o-cresol. Some long aromatic species such as biphenyls also appear in this fraction. Compared to fraction #8, the phenols in fraction #9 are of shorter size while the peaks appearing at long GC retention times are aromatics. It is safe to say that phenols appear before 16 minutes of GC retention time followed by aromatics after 16 minutes. The GC of fraction #10 is shown in Figure 5-7. Light phenols including xylanols and cresols present in this fraction are separated on the GC before a retention time of 8 minutes. The species appearing after 8 minutes are aromatics, mostly with alkyl side chains.

Fraction #11, whose GC is in Figure 5-8, contains phenol, which appears at 3 minutes. Phenol is the only phenolic in Fraction 11. Almost all possible isomers of one and two ring aromatics with alkyl side chains (propyl or shorter) are detected in this fraction. Since the number of species are higher, co-elutions of two or more components at one GC retention time is observed. The Mass Spectral fragmentation pattern can be used to assign the molecular formula and general structural nature. The identification of isomers is very difficult in a number of cases. The NBS Mass Spectral Data Base has only a fraction of the needed standard reference spectra to identify the species in this fraction. Most of the identification has been assigned based on the fragmentation patterns and boiling points derived from the GC retention times.

Fraction #12 as shown in Figure 5-9 has overlapping from two types of aromatics - alkylated aromatics and polycyclic aromatics.

Fraction #13 contains aromatics with slight alkylation and the ring numbers increase as shown in Figure 5-10. Both fractions #14 and #15 (Figure 5-11, 12) contain only aromatics with few alkyl side chains.

One exception to the rule that SEC separates species in decreasing order of linear molecular size is that condensed ring aromatics tend to remain in the column longer. Some polycyclic aromatics such as pyrene and coronene are eluted from the gel column only after napthalene although they are much larger. More pyrene is in Fraction #15 than in Fraction #14 but the reverse is true for anthracene which appears before naphthalene.

Effect of Solvent-Solute Interaction on SEC

The separation of chemical species by size exclusion chromatography is more reproducible than any other type of chromatography. Once the SEC columns, the mobile phase (most often a pure solvent like THF or toluene), and the flow rate are selected, the retention volume (or retention time assuming the flow rate does not change) is primarily a function of linear molecular size, which can be obtained from the valence bond structure if the compound is known. Some of the chemical species can interact with the solvent forming complexes with an effective linear size greater than that of the molecule. This causes the expected retention volume, based on "free" molecular structure, to shift to a lower but very reproducible retention volume. Phenols in coal liquids form 1:1 complex with THF (9,10) and carry the effective linear molecular size to increase. As a result phenolic species elute sooner than expected from their

valence bond structures. Although phenols and aromatics have
similar structures, they elute at different retention volumes
when THF is used as the mobile phase.

Effect of Gel-Solute Interaction on SEC

If the molecules have a tendency to interact with the packing
material of the column, molecules may elute at longer retention
volume. In the case of PL gel columns which are packed with a
gel formed by the co-polymerization of styrene and divinyl
benzene, the aromatic species have a tendency to stay on the
column slightly longer than expected from their linear molecular
sizes. The SEC elution pattern shows that the aromatics appear
to be smaller than similar structured cycloalkanes. Among
similar aromatics such as benzene, naphthalene, and anthracene,
the elution volume is in the decreasing order of linear molecular
sizes (i.e. anthracene is followed by naphthalene and then
benzene elutes last). In the case of polycyclic aromatics where
three or more rings are attached to a single ring as in pyrene
and coronene, the elution time increases as the number of rings
increases. Pyrene has a retention time longer than that of
anthracene. Coronene is eluted only after pyrene. The
additional alkyl side-chain causes the molecules to elute sooner,
as expected from the resulting linear size increase due to alkyl
side-chains.

Probable Molecular Structure Based on SEC

Although the mechanism of SEC separation is controlled by linear
molecular size as well as other parameters, the separation
pattern is very reproducible. Considering all the molecular
parameters responsible for the size exclusion chromatographic
separation pattern and the known separation patterns of a number
of compounds, it is possible to predict the retention volume of a
compound of known strucure. Based on the same principle the
retention volume gives information on the structure of the
molecule.
    The role of size exclusion chromatography is the separation
of rather complex coal liquids into simpler fractions. The retention volume can be used to help identify the chemical structure
where GC-MS is unable to identify its possible structure. For
example biphenyl and dihydroacenaphthene have the same molecular
formula as well as similar mass spectral fragmentation patterns.
Coal liquids contain both species. The one which appears first
(lower SEC retention volume) is biphenyl (GC ret. time ≈ 17 min.
in Figure 5-6). Dihydroacenaphthene appears later at longer SEC
retention volume and is identified in Figure 5-12 at GC retention
time of 13 minutes. The former has a longer structure compared
to the latter.

Sample Spreading

As mentioned earlier, SEC is extremely reproducible with respect
to retention volume, peak width, and height. The ratio of reten-

tion volume to peak width (V/W) is a constant over the separation range of the SEC column. The separation range is between total size exclusion volume (V_E), where all the species larger than a particular size elute without any size separation, and total permeation volume (V_D), where all the molecules smaller than a particular size elute without any separation irrespective of their sizes. The species which fall within the size range of V_E to V_D, which is determined by the pore size of the column, are eluted somewhere between V_E and V_D in decreasing order of linear molecular sizes. The peak width increases as the retention volume increases.

This phenomenon is quite apparent from Figure 5. Since the 100 µl fractions are collected at 0.5 ml volume intervals and the species have about 1 µl peak width, each species is detected in 2 or more consecutive fractions. The fractions at the lower retention volumes spread less compared to those at the higher retention volumes, as illustrated by examining the spreading of $C_{25}H_{52}$ and phenathrene. The compound $nC_{25}H_{52}$ is detected in three SEC fractions with a maximum concentration in the second fraction (Figure 5 - 1, 2, 3). Phenanthrene is detected in the last five fractions (Figures 5 - 8, 9, 10, 11, 12). The examination of any particular species shows that they are spread over 3 or more fractions and the range of spreading (a measure of peak width) increases with retention volume.

It is quite evident that the analysis of small fractions at retention volume intervals of 0.5 ml or even 1 ml does not miss any major or minor species. Almost all species at detectable levels are identified.

Quantitative Analysis by SEC-GC-MS

The quantitative estimation of species by SEC-GC-MS technique requires a mathematical solution. Two types of approaches for the quantitative estimation can be envisioned. One for the estimation of one or more selected species of interest. The second approach is based on grouping of various species in coal liquids into a few chemical lumps and estimating the quantity of these lumps by using the data derived from the analysis is technique.

When a particular component eluting at a certain retention volume is to be estimated, this approach can be outlined as follows. Since SEC is extremely reproducible, the peak shape, peak width and peak height are dependent on the amount of the species in the sample volume injected, sample volume and retention time. From these factors the SEC peaks can be simulated or elution pattern of any species within the separation range can be plotted as a function of mass vs. retention volume. The analysis data supplies the concentration of this particular species over two or more 0.5 ml intervals. A match-up computer program has to be developed so that it can pick up the peak shape and concentration based on 3 or 4 data points at known intervals.

The second approach towards quantitative analysis is based on dividing the coal liquid into distinct fractions containing simi-lar chemical species as is illustrated in Figure 2. This type of

chemical lumping gives more useful information on coal liquefaction reactions and on kinetic models of coal liquefaction processes.

It is a fact that coal liquids are composed of thousands of
or maybe millions of chemical compounds. Classifying them into a
few meaningful groups of compounds can be a suitable procedure to
evaluate coal liquid compostion. Currently at least two types of
classification approaches have been used. One is based on physical separation by appropriate combination of physical and
chemical methods. The second approach is based on estimating
functional groups or functionalities in a coal liquid by sophisticated instrumentation such as solid state NMRs and FT-IRs. By
the first methodology coal liquid is separated into fractions
such as oils (pentane soluble) asphaltenes (pentane insolubles
but benzene solubles) and preasphaltenes (pyridene soluble but
insoluble in benzene and pentane) where no clean separation or
estimation is achieved. By the second type of methodology more
cummulative data is achieved on functional groups. This methodology has the disadvantage of looking at a compound with more than
one functional group. The amount of each group is estimated as
separate moieties and computed separately. The alkyl chains
attached to an aromatic ring, which also has phenolic groups, is
very different from saturated hydrocarbons such as normal
paraffins but they are classified together in this approach.

Distribution of Alkanes

One of the major results of SEC-GC-MS studies is the discovery of
an orderly pattern, by which various isomers and homologs of
similar chemical species exist in coal liquids. For example
almost any direct coal liquefaction process produces very similar
species, which differ from each other by size and extent of isomerization but with an orderly distribution pattern. Alkanes
ranging from $C_{12}H_{26}$ and $C_{44}H_{80}$ are detected in almost any coal
liquid. Most of these are straight chain alkanes showing an
orderly continuous pattern. Neither is a particular n-alkane
almost absent nor is it present in a disproportionate amount.
Exceptions exist for some branched alkanes such as pristane,
phytane, and hopane. These species are also called biomarkers
and their concentration varies depending on the sample.

Distribution of Phenols

Phenols are a major chemical lump present in coal liquids.
Phenols have basically one or more aromatic ring structures with
alkyl substituents. Methyl, ethyl and propyl are the most common
alkyl substituents. The smallest specie is the one with a
hydroxyl group attached to a benzene ring. Addition of a methyl
group produces three isomers - o-, m-, and p-cresols. It appears
that all three are present in more or less same proportion. The
number of possible isomers increases as the possible number and
size of alkyl substituents increases. It is expected that higher

degree of alkylation can produce larger molecules in a larger
number of isomeric forms, separation of which is rather difficult
even by high resolution GC methods. This could be the reason why
a shift in the GC base line is observed for the SEC phenolic
fractions rather than resolved peaks. Since these shifts are
quite reproducible and real, it can be assumed that these "bumps"
are due to a large number of components eluting continuously
without resolution. Their SEC retention time suggests that they
are probably phenols. The gas chromatographers who are used to
fewer sharp peaks from capillary GC may prefer to resolve them.
Sometimes derivitization techniques are used to obtain sharper
well resolved peaks. As a matter of fact unresolved "bumps" are
telling a story. Too many isomers of close molecular weight or
boiling point are eluting without resolution at close retention
times. Phenols do show peak tailing in most GC separation conditions. But currently available capillary columns do not have
this as a serious problem. Peak tailing is expected to decrease
as the degree of alkylation increases. Peak tailing for cresol
is less than that for phenol. It is much improved for xylanols.
The derivitization of phenols prior to GC separation may produce
fewer well resolved peaks but at the expense of losing some
components.

Distribution of Aromatics

The number of isomers of alkylated aromatics is enormous. Lower
members of alkylated benzenes such as xylenes are well resolved
and detected by FID and MS. Increased alkylation causes an increase in the number of isomers. In the case of both alkylated
phenols and aromatics various isomers are existing in a
continuous pattern. The lower alkylation gives few well resolved
isomers. The higher alkylation gives a large number of isomers
but in smaller concentrations.

Overlapping of Species in SEC Fractions

When SEC-GC is used for coal liquid analysis, 0.1 µl fractions of
SEC effluents are analyzed by GC to produce simpler gas chromatograms. Some of these gas chromatograms, for example the GC of
longer alkanes, are composed of chemically similar components.
The flame ionization detector (FID) response factor based on
mass, is essentially the same for these larger alkanes. The
total area counts of such gas chromatograms, excluding solvent
peak, which represents the sample volume (0.1µl), multiplied by
the response factor will give the mass of alkanes in the SEC
fraction analyzed. Certain SEC fractions are composed of two or
more different chemical species due to the overlapping effect of
similar size species. For example, the low boiling point alkanes
are mixed with the high boiling point phenols where the linear
molecular sizes of the species are similar. The alkanes appear
at low retention times whereas phenols appear at longer retention
times (Figure 5-4). In these cases the area counts have to be
lumped into two groups, one for alkanes and another one for
phenols. Each of these area counts multiplied by the

corresponding FID response factor indicate the amount of alkanes or phenols present in the 0.1μl SEC fractions. All of the sixteen or more GCs of selected SEC 0.1μl fractions of coal liquids or recycle solvents are individually analyzed for various "lumped chemical" species in the fractions. Coal liquid samples can be separated by distillation and estimated for the nonvolatile content. The SEC of nonvolatiles and volatiles are reconstructed to show both in the same SEC output. These data along with SEC-GC data are used to reconstruct the SEC of Wyodak coal derived recycle solvent as shown in Figure 4.

SEC-GC-MS Analysis of Other Coal Liquids

SEC output of four coal liquids are shown in Figure 6. The gas chromatograms of SEC fractions collected at similar retention volumes contain similar chemical species as illustrated in Figure 7. The amount of each component may vary from sample to sample depending on the coal liquid. Wyodak recycle solvent has less pristane, a biological marker compared to other three samples.

SEC vs. Distillation

The chemical lumping pattern shown in Figure 4 is very similar to the plotting of distillation temperatures vs. composition, a technique commonly used in petroleum refining to simulate the composition of distillate as a function of temperature. Since SEC includes nonvolatiles, information on their size distribution is also shown. In each chemical lump the molecular weight decreases as SEC retention volume increases. The individual chemical lump has a SEC separation pattern similar to a distillation temperature vs. molecular weight plot, a technique used in petroleum refining to illustrate the composition of various distillation cuts.

CONCLUSIONS

The chemical lump of alkanes is the simplest. Straight alkanes are distributed throughout the range in a continuous pattern. Such a pattern does not exist for phenols and aromatics. Both of them have similar aromatic nucleus such as benzene, indan and naphthalene. The presence of hydroxyl groups distinguishes the phenols from the aromatics. The alkyl side chains ranging mostly from C_1 to C_3 attached to the simple aromatic nucleus result in larger molecules of phenols and aromatics. As the number of side chains increases, the number of isomers increases exponentially. The mass distribution of phenolics and aromatics reaches a maximum at a certain molecular weight and then decreases at higher molecular weights. At the lower end of the mass distribution pattern, since a number of isomers are possibly smaller, the GC is well resolved and composed of larger well resolved peaks. At the higher molecular weight end, a very large number of isomers are possible in a small mass, the GC shows an upward shift in the base line which is due to a large number of species that are appearing

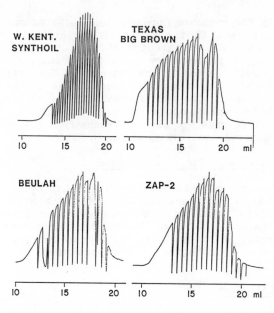

Figure 6.   SEC of coal liquids using the SEC-GC interface on-
line (Figure 2).   Fraction collection timing is
similar to one used for Wyodak (Figure 4) except for
Western Kentucky Synthoil where 24 fractions were
collected.

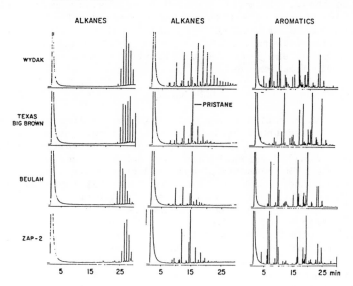

Figure 7.   Gas chromatograms of SEC fractions, collected at
similar retention volumes, of four recycle solvents.

unresolved. Since phenols have an inherent tendency for tailing, the large phenols are not at all resolved. In the case of large aromatics, the enormous number of isomers are appearing with quite a few peaks partially resolved.

The species which are unknown and have not been identified as one of the major chemical lump such as alkanes, phenols and aromatics are lumped together as unidentified. However, the species in this lump include saturated and unsaturated cycloalkanes with or without side chains, which resembles the naphthenes, a petroleum refinery product group. A number of well known species in coal liquid are not mentioned in this lumping scheme. Such as heterocyclic compounds with sulfur, nitrogen or oxygen as the heteroatom, and other heteroatom containing species. Some of these compounds appear with aromatics (e.g. thiophenes, quinolines) and with phenols (e.g. aromatic amines), and most of them are lumped with the unidentified species lump.

One exception to the rule that SEC separate species in the decreasing order of linear molecular size, is that condensed ring aromatics appear to adhere to the column longer so that some polycyclic aromatics such as pyrene and coronene are eluted from the gel column only after naphthalene although their molecular sizes are much larger. More pyrene is in Fraction #15 than in Fraction #14 but reverse is true for anthracene.

Acknowledgments

The financial support of the U.S. Department of Energy, (project Number DE-AC18-83FC10601), Texas A&M University Center for Energy and Mineral Resources and Texas Engineering Experiment Station is gratefully acknowledged. The Energy Research Center at University of North Dakota furnished samples for the study.

Disclaimer: This report was prepared as an account of work sponsored by an agency of the United States Government. Neither the United States Government nor any Agency thereof, nor any of their employees, makes any warranty, or assumes any legal liability.

Literature Cited

1. Hendrickson, J.G., **Anal. Chem.**, 1968, **40**, 49.
2. Majors, R.E.J., **Chromatog. Sci.**, 1980, **18**, 488.
3. Cazes, J. and Gaskill, D.R., Sep. Sci., 1969 4, 15.
4. Krishen, A. and Tucker, R.G., **Anal. Chem.**, 1977 **49**, 898.
5. Philip, C.V., Anthony, R.G., **Fuel Processing Technology**, 1980, **3**, 285.
6. Zingaro, R.A., Philip, C.V., Anthony, R.G., Vindiola, A., **Fuel Processing Technology**, 1981, **4**, 169.
7. Philip, C.V., Zingaro, R.A., Anthony, R.G. in "Upgrading of Coal Liquids," Ed. Sullivan, R.F., **ACS Symposium Series No. 156**, 1981;p.239.
8. Philip, C.V., Anthony, R.G., **Fuel**, 1982, **61**, 351.
9. Philip, C.V., Anthony, R.G., **Fuel**, 1982, **61**, 357.

10. Philip, C.V. and Anthony, R.G., "Size Exclusion Chromatography", Ed. Prouder, T., **ACS Symp. Series**, 1984, **245**, 257.
11. Philip, C.V., Anthony, R.G. and Cui, Z.D., "Chemistry of Low-Rank Coals, Ed. Schobert, H.H., **ACS Symp. Series**, 1984, **264**, 287.
12. Philip, C.V., Bullin, J.A. and Anthony, R.G., **Fuel Processing Technology**, 1984, **9**, 189.
13. Sheu, Y.H.E., Philip, C.V., Anthony, R.G. and Soltes, E.J., **Chromatographic Science**, 1984, **22**, 497.
14. Philip, C.V. and Anthony, R.G. **Am. Chem. Soc. Div. Fuel Chem. Preprints**, 1985, **30**, (1), 147.
15. Philip, C.V. and Anthony, R.G. **Am. Chem. Soc. Div. Fuel. Chem. Preprints**, 1985, **30** (4), 58.

RECEIVED May 21, 1987

DATA ANALYSIS

Chapter 12

Chemometrics in Size Exclusion Chromatography

Stephen T. Balke

Department of Chemical Engineering and Applied Chemistry, University of Toronto, Toronto, Ontario M5S 1A4, Canada

Nonlinear regression, graphics and error propagation analysis are reviewed along with their applications in size exclusion chromatography (SEC). Nonlinear regression is the fitting of equations which are non-linear in the unknown coefficients, to experimental data. The method of implementation depends upon the results required. Simple data fitting is distinguished from obtaining meaningful coefficient values in fitted equations. Calibration curves from polydisperse samples, calibration curve fits, detector nonlinearity assessment and shape function fitting are discussed. Graphics portray data for assessment and new insights. Plotting of digitized chromatogram heights, residuals, and integrands (as "moment analysis plots") are examined. Current common but misleading plots are described. Error propagation analysis provides a simple way of predicting error in calculated quantities. Ratioing experimental quantities is shown to be a source of large error.

Chemometrics is computer implemented mathematics, particularly statistics, in Chemistry. With the new wave of microprocessor assisted liquid chromatographs and inexpensive microcomputers, chemometric methods in column liquid chromatography are now rapidly developing and proliferating.

Size exclusion chromatography (SEC) has traditionally required more application of computers than other forms of liquid chromatography. For example, in high performance liquid chromatography (HPLC), only an area determination under individual, separated, chromatographic peaks is generally required. This area is needed for determining the concentration of the single solute present under each peak. By contrast, in SEC the conventional objectives of the interpretation are to calculate the concentration of each of thousands of different molecular weights present under one peak (the molecular weight distribution) and to calculate the molecular weight averages.

0097–6156/87/0352–0202$06.00/0
© 1987 American Chemical Society

In this chapter, three chemometric methods of increasing importance to SEC are examined: nonlinear regression, graphics and error propagation analysis. These three methods are briefly described with emphasis on SEC applications and on critical concerns in their correct implementation. In addition to the specific references cited, further information on these methods and others may be found in a recent book which examines chemometrics in both SEC and HPLC together (1) as well as in periodic reviews (2).

Nonlinear Regression: Description

Regression generally means the fitting of mathematical equations to experimental data (3). Nonlinear regression, unlike linear regression, encompasses methods which are not limited to fitting equations linear in the coefficients (e.g. simple polynomial forms).

Nonlinear regression methods are extremely simple and flexible. They are composed of three parts:

1. A "search algorithm" which successively guesses values for the unknown "coefficients" (these are also called "parameters" or "constants").
   The Nelder-Mead SIMPLEX algorithm has been frequently used in Analytical Chemistry as well as in other areas of science and engineering. Assessment and further development of the method remains an active field of research (4).

2. An "objective function" which defines the fit of the equation to the data given the guessed coefficient values.
   Residuals are the difference between the value of the ordinate of the experimental point and the value of the corresponding ordinate given by the fitted equation. That is, if data is fitted as y versus x, then at any x_i the residual is the difference between the data value of y_i and the value on the fitted line ($\hat{y}_i$):

$$\Delta_i(y) = y_i - \hat{y}_i \tag{1}$$

Simple linear regression (3) actually minimizes the sum of squares of the residuals in order to fit an equation to data. That is:

$$\text{minimize} \sum_{i=1}^{n} \left(\Delta_i(y) \right)^2 \tag{2}$$

where n is the number of data points.

Nonlinear regression can employ the same "objective function". However, nonlinear regression is much more flexible. The desire of the analyst to emphasize some data over another for experimental error reasons or for other reasons can be easily incorporated via weighting factors, powers higher than squared, or even novel functional forms. For example, a general nonlinear regression objective function similar to that of Bandler (5) is:

$$\text{minimize} \sum_{i=1}^{n} w_i \left| \Delta_i(y) \right|^p \tag{3}$$

where the w_i are weighting factors and p is a constant.

3. A method of constraining the guesses of the search to physically reasonable values.

Two primary methods exist for constraining the search to values considered reasonable by the analyst: transformation of the independent variables and penalty functions (5). The main point to note is that with nonlinear regression the analyst should have this very significant power over the search. Then, instead of trying to find a solution from an infinite number of possible values for the unknown coefficients, the search will focus on a physically reasonable range.

Regression methods are now frequently employed in SEC interpretation and it is important to realize that the validity of a particular application strongly depends upon the purpose of the regression. There are three main reasons for applying regression methods in column liquid chromatography (1):

o To summarize the data so that it can be readily
   regenerated and "in between values" accurately
   determined.
   This is the easiest type of application. Only
   precision and accuracy of the fit are of concern.

o To determine the value of physically meaningful
   coefficients in equations.
   This is the most difficult application. Not only
   must the fit be accurate and precise, the precision
   of the coefficients determined in the fit must be
   defined.
   Correlation amongst the coefficients and sensitivity
   of the predictions of the equation used to the values
   of the coefficients is also important.

o To provide efficient logic on how to change
   separation conditions in order to obtain optimum
   resolution.
   This application has been used in HPLC but so far not
   in SEC (6). Since reduction of experimental
   work is the emphasis, the efficiency of the fitting
   method is of utmost importance.

Nonlinear Regression: Examples in SEC

Determining Calibration Curves from Polydisperse Samples.   In
conventional SEC interpretation, narrow molecular weight distribution
standards are needed for calibration purposes.   Nonlinear regression
has enabled polydisperse samples to be used.   A variety of methods
are involved.   In the simplest one, developed in 1969 (7), a
polydisperse standard of the polymer of interest with known $\overline{Mn}$ and $\overline{Mw}$
is available.   It is injected into the SEC to obtain a broad
chromatogram.   A calibration curve equation containing two unknown
coefficients is assumed (e.g. a "linear " calibration curve) and the
nonlinear regression method is used to find the coefficients such
that when the calibration curve is used along with the broad
chromatogram to calculate molecular weight averages $\overline{Mn}$ and $\overline{Mw}$, these
averages agree with the independently known values.   The same
calibration curve is then used to analyze other similar samples of
unknown $\overline{Mn}$ and $\overline{Mw}$.   There are now a multitude of "calibration curve
search methods" with new ones regularly appearing in the literature.
For example, some employ resolution correction expressions as well as
a calibration curve equation.   The schematic in Figure 1 shows one
such strategy: the molecular weight averages are corrected for axial
dispersion effects ($\overline{Mn}(c)$ and $\overline{Mw}(c)$) before being compared to those
known for the standard ($\overline{Mn}(a)$ and $\overline{Mw}(a)$).   As shown in Figure 1,
intrinsic viscosity of the whole polymer can take the place of one
molecular weight average.   Then, $[\eta](c)$ is the resolution corrected
value of intrinsic viscosity and $[\eta](a)$ is the value known for the
standard.   Sometimes the resolution correction factors themselves
constitute some of the unknown coefficients which are to be
determined by the search.   Others search methods utilize the
universal calibration curve rather than the conventional one and
intrinsic viscosity (expressed as a function of molecular weight) as
well as molecular weight averages.   This provides one way of
calibrating for branched polymers.   Recent reviews have been
published (1,8,9).   As mentioned above, if the actual coefficient
values are to be allocated some physical significance (e.g. as
resolution factors) then additional considerations beyond those of
ensuring a simple fit to the data enter the problem.

Fitting of Calibration Curves Determined Using "Monodisperse"
Samples.   The mathematical fit of the calibration curve of log
(molecular weight) versus retention time (or for universal
calibration, log (hydrodynamic volume) versus retention time)
determined in conventional interpretation using narrow molecular
weight distribution standards drastically affects computed results.
The calibration curve is used in calculating both ordinate and
abscissa of the molecular weight distribution as well as the
molecular weight averages.   Fitting the plotted points "by eye" is
not sufficiently dependable.   Frequently, linear regression is used
because a simple polynomial form is assumed.   The equation is then
linear in the unknown coefficients.
      Equation 4 is an example of such an equation:

$$\log M = B + C\,t + D\,t^2 + E\,t^3 \qquad\qquad (4)$$

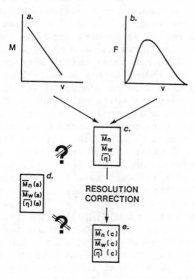

Figure 1: Schematic diagram of calibration curve search with no resolution correction (a,b,c,d) and with direct resolution correction of the molecular weight averages (a,b,c,e,d). (Reproduced with permission from Ref. 1. Copyright 1984, Elsevier.)

M is the peak molecular weight of a narrow distribution linear homopolymer and t is retention time. The equation is linear in the coefficients B, C, D and E.

Note that, as shown by Equation 4, because the equation can be fit by linear regression does not necessarily mean that it is a straight line. Linearity in the coefficients is the important aspect, not linearity in the dependent variable, t. However, as will be seen in the discussion on error propagation analysis below, because "log M" and not simply "M" must be used in the linear regression, it does mean in this case that each peak molecular weight must have the same percent error. If not, then nonlinear regression or specifically derived weighting factors in the linear regression must be used.

The most common reason for using nonlinear regression however, is that the equation of interest is nonlinear in the coefficients to be determined. For example, in their SEC interpretation, Provder and Rosen (10, 11) as well as Cardenas and O'Driscoll (12) used an expression nonlinear in the coefficients B,C,D and E, derived by Yau and Malone (13) in a theoretical development:

$$v = B + C \left[\frac{1}{\sqrt{\pi} \, z} \, (1-\exp(-z^2)) + \text{erfc}(z) \right] \qquad (5)$$

where

$$z = \frac{M^E}{D}$$

and

$$\text{erfc}(z) = \frac{2}{\sqrt{\pi}} \int_z^\infty \exp(-x^2) \, dx$$

where v is retention volume.

Another example of an equation nonlinear in the parameters is the sum of exponentials form which is useful with some resolution correction methods:

$$M = B \exp(-Ct) + D \exp(-Et) \qquad (6)$$

Equation 6 is nonlinear in the coefficients B, C, D and E.

Detector Nonlinearity Assessment. Detector linearity is implicitly assumed when conventional chromatogram interpretation is accomplished. This assumption must be checked by plotting the the area under a narrow chromatogram versus concentration for various concentrations and fitting the data with a polynomial to see if Beer's law holds. That is, the polynomial:

$$A = B + Cc + Dc^2 + Ec^3 \qquad (7)$$

where A is area, c is concentration and B,C, D and E are coefficients, is fit to the data using linear regression. If c is

the only non-zero coefficient then Beer's law holds. Otherwise,
either more terms in the polynomial are retained or alternatively,
the curve can be characterized by an equation nonlinear in the
coefficients and nonlinear regression used to fit the equation to the
data. For example, Scott (14) has suggested an equation of the form:

$$A = K c^r \tag{8}$$

where K and r are the coefficients to be determined. This equation
can be linearized by using logarithms with analogous conditions for
error in log A to those mentioned above for log M versus t fitting.
Also, Carr (15) has criticized Equation 8 on the basis that it cannot
adequately fit certain cases.
Fitting of Shape Functions. Shape functions are mathematical
expressions which are used to describe the chromatogram of a sample
which has only one molecular size present. Shape functions are
needed to quantitatively express the amount of "band broadening" or
"peak spreading" which occurs in SEC due to axial mixing effects in
the chromatographic columns. These functions can be complex
expressions containing several unknown coefficients. They are fit to
chromatograms of "single molecular weight" samples or are
incorporated into more complex expressions to fit broad
chromatograms. Even the simplest and most common, the Gaussian
function, is a function nonlinear in three coefficients, area (A),
standard deviation (σ) and mean time (t):

$$G(t) = \frac{1}{\sqrt{2\pi}} \frac{A}{\sigma} \exp\left\{ \frac{-(t - \bar{t})^2}{2\sigma^2} \right\} \tag{9}$$

Figure 2 shows a fit of this function to the front half of an
experimental chromatogram.

Resolution Correction. As alluded to above, nonlinear regression is
often employed for resolution correction in approaches involving
combined calibration curve/resolution correction search methods, and
in determining "resolution factors" (e.g. the reciprocal of the
variance in the Gaussian Function (Equation 9). Note that when the
coefficient in an equation, such as the variance, is assigned
specific meaning (e.g. a measure of resolution) then the regression
is being used beyond a simple "data regeneration" purpose. The
second of the main reasons for applying regression, namely the
determination of the value of physically meaningful coefficients, is
then involved, with its attendant additional concerns.

Graphics: Description

Graphics are simply displays of plotted data used to assess the
validity of calculations and to direct interpretation. With the
revolutionary increase in computer accessibility, graphics have
become very easy to obtain. Graphics enable the extremely rapid and

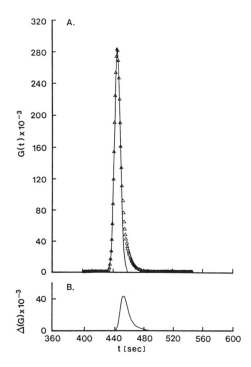

Figure 2: A. Fit of a Gaussian shape function to the front half of an experimental chromatogram. B. Residual between the Gaussian and the experimental curve. (Reproduced with permission from Ref. 1. Copyright 1984, Elsevier.)

clear communication to the chromatographer of many diverse diagnostic messages simultaneously. There is a wide variety of possibilities. Those most important to SEC are reviewed in the next section.

Graphics: Examples in SEC

Plotting of Digitized Chromatogram Heights. "Overlapping" of chromatograms has been used by chromatographers since the earliest days of SEC. Allowing the computer to "normalize" the chromatograms by dividing each height by the area enables overlapping to be easier interpreted because then all the chromatograms have an area of unity. Multiplying each normalized height by the amount of polymer present in a polymerization process so that the resulting area of the chromatogram reflects polymer quantity has proven very useful in at least one investigation (16). Furthermore, simply plotting of the actual points obtained by the computer and a knowledge of how calculations are being performed can provide an immediate check on data sampling rate. Figure 3 shows digitized heights plotted with a continuous curve superimposed on the points. If it is known that the computer is drawing a straight line between data points in the various integration operations then we can see that because of the inadequate sampling rate, the peak used by the computer in this case is a truncated version of the more probable curve shape.

Plots of Residuals. Residuals can be plotted in many ways: overall against a linear scale; versus time that the observations were made; versus fitted values; versus any independent variable (3). In every case, an adequate fit provides a uniform, random scatter of points. The appearance of any systematic trend warns of error in the fitting method. Figures 4 and 5 shows a plot of area versus concentration and the associated plot of residuals. Also, the lower part of Figure 2 shows a plot of residuals (as a continuous line because of the large number of points) for the fit of the Gaussian shape to the front half of the experimental peak. In addition to these examples, plots of residuals have been used in SEC to examine shape changes in consecutive uv spectra from a diode array uv/vis spectrophotometer attached to an SEC and the adequacy of linear calibration curve fits (1).

Moment Analysis Plots. Moment analysis plots are obtained by plotting the integrand of the integral defining a moment of a chromatogram or a molecular weight average (which is really the moment of a transformed chromatogram). For example, a moment analysis plot for $\overline{Mw}$ is based on the definition of $\overline{Mw}$ in terms of the chromatogram:

$$\overline{Mw} = \int_{0}^{\infty} M(t)\ W_N(t)\ dt \qquad (10)$$

where:

$M(t)$ is the molecular weight at retention time t
$W_N(t)$ is the normalized height of the chromatogram ("normalized" means that each height on the raw chromatogram has been divided by the area).

420   440   460   480

t [sec]

Figure 3: Assessment of sampling rate adequacy by plotting of
digitized heights. (Reproduced with permission from Ref. 1.
Copyright 1984, Elsevier.)

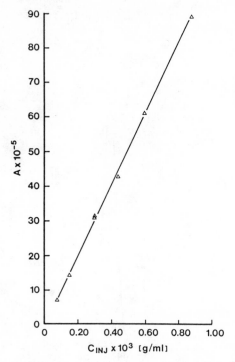

Figure 4: Linear regression of peak area versus injected solute concentration. (Reproduced with permission from Ref. 1. Copyright 1984, Elsevier.)

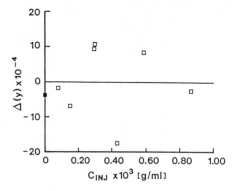

Figure 5: Plot of residuals for the linear regression of Figure 4. (Reproduced with permission from Ref. 1. Copyright 1984, Elsevier.)

The product $M(t)$ $W_N(t)$ is the integrand in this case and a plot of $M(t)$ $W_N(t)$ versus t is the moment analysis plot for $\overline{Mw}$. Similar plots can be made for the other molecular weight averages. Figure 6 shows moment analysis plots for $\overline{Mn}$, $\overline{Mw}$ and $\overline{Mz}$.

Moment analysis plots to provide a picture of the calculation of a molecular weight average were first proposed by Boni (17). When superimposed on the original chromatogram, this picture enables the analyst to see just what heights of the original chromatogram are contributing to a molecular weight average. Sometimes, molecular weight averages are seen to depend upon heights at the noise level of the detector for most of their information. In Figure 6 we see some evidence of this in the low retention time portion of the moment analysis plot for $\overline{Mz}$.

Misleading Plots. Although graphics is now sufficiently important to be considered a separate chemometric method it is important to realize that highly misleading plots can readily be obtained. In particular, plotting the same variable, even as part of a group of other variables, on both axes of a plot can create a correlation (the same as plotting "x" versus "x" for example) which has no physical significance. Using logarithms on both axes can reinforce this misleading appearance (18). This error is particularly liable to affect the non-SEC characterization of molecular weight standards used for SEC calibration. The reason for this is that in interpretation of light scattering, intrinsic viscosity and osmometry data, the customary procedure is to plot a property value (e.g. specific viscosity) divided by concentration versus concentration and to extrapolate the apparent straight line to zero concentration. Garcia-Rubio et al. (19) show that such plots can provide a curvature which falsely indicates the need for more terms in the fitting equation (i.e. more "virial coefficients").

Error Propagation Analysis: Description

Error propagation analysis is the estimation of error accumulation in a final result as a consequence of error in the individual components used to obtain the result. Given an equation explicitly expressing a result, the error propagation equation can be used to estimate the error in the result as a function of error in the other variables. For example, the error in a result "z" which is a function of two variables, "x" and "y", each of which is subject to random error of standard deviation s_x and s_y respectively where the random error in one variable is not correlated to the random error in the other, is given by (20, 21):

$$s_z^2 = \left[\frac{\partial z}{\partial x}\right]^2 s_x^2 + \left[\frac{\partial z}{\partial y}\right]^2 s_y^2 \qquad (11)$$

where s_z^2 is the error variance of the result, z.

For variables whose errors are not independent, a general form of Equation 11 incorporates covariance as well as variance. Also, another form of the equation has been derived for non-random error (20). As will be seen below, certain computations, well known in SEC, are now being found to be intrinsically imprecise because of error propagation.

Error Propagation Analysis: Examples in SEC

Estimation of Error in Calculated Results. A startling result is obtained when the error propagation equation for random error is applied to a result which is calculated from the ratio of two experimental quantities, each with random experimental error. If the components of the ratio are independent variables then it is easily shown that the error in the result increases at lower values of the denominator of the ratio despite highly precise experimental data (see (22) and Appendix I). The effect of this on extinction coefficient values computed using ratioing and needed for SEC analysis of copolymers is serious (22). Figure 7 shows the calculated extinction coefficient values with different symbols for alternating, random and block copolymers. Figure 8 shows the same values with error bars obtained from the error propagation analysis of Appendix I. Differences between the various types of polymers no longer appear significant (22).

Even when there is some dependence amongst the experimental variables, error propagation can mean that ratioing of experimental values causes unexpectedly high error. Work in analysis of the "internal standard method" commonly used in chromatography (mostly HPLC) is one example (23). Absorbance ratioing, an increasingly common approach (24), is subject to similar possibilities.

Error in Calculated Data to be Fit by Regression. Error propagation can be used to estimate the error in calculated data and plotted along with the data as error bars before equations are fit. Linear regression, for example, requires that the error be the same for each ordinate value and that no significant error be present in the abscissa values. Because the logarithm of molecular weight is used in the ordinate it can be shown by error propagation analysis that, as mentioned above, application of linear regression to the conventional calibration curve really means that the "percent" error in molecular weight is assumed constant (see Appendix II). For molecular weight this is fortunate since a constant "percent" error usually is the case. However, for many other cases, the best procedure is to avoid the transformation by using nonlinear regression.

Conclusions

o Nonlinear regression enables broad molecular weight distribution standards to be used for SEC calibration and permits simultaneous resolution correction and calibration. Furthermore, it greatly increases the variety of equations which can be fit to calibration curves, detector linearity data and chromatogram shapes.

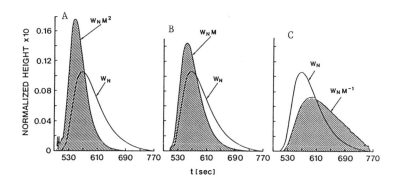

Figure 6: Moment analysis plots for molecular weight averages: A. $\overline{Mz}$, B. $\overline{Mw}$, C. $\overline{Mn}$. All curves have been normalized to give unit areas by dividing each height by the respective area. Compare what fraction of shaded area corresponds to the very small tail heights of the normalized chromatogram (i.e. W_N) for each plot. (Reproduced with permission from Ref. 1. Copyright 1984, Elsevier.)

Extinction Coefficient

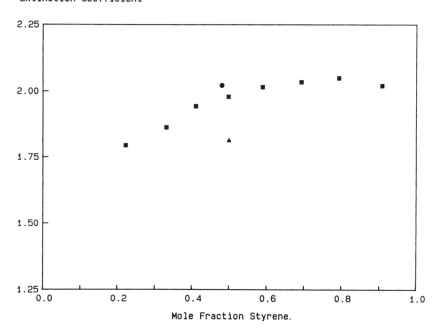

Figure 7: Extinction coefficients calculated from ratio of absorbance to concentration versus mole fraction styrene: ▲ alternating copolymer; ■ random copolymer; ● block copolymer. (Data analysis from Ref. 22.)

Extinction Coefficient

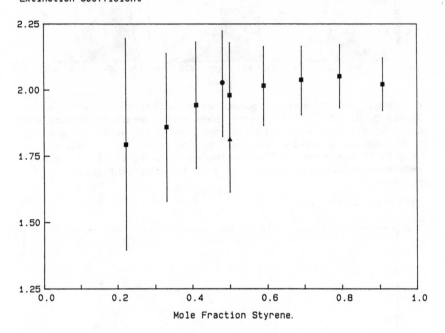

Mole Fraction Styrene.

Figure 8: Same as Figure 7 except showing 95% confidence limits as vertical error bars about each point. (Data analysis from Ref. 22.)

o For valid results, the nonlinear regression procedure
   used must fit the application. Interpolation,
   determination of physically meaningful coefficient
   values and experimental optimization, constitute the
   three main applications.

o Graphics is now sufficiently important to constitute a
   separate chemometric method. It is the key to rapid
   diagnosis of numerical interpretation problems by
   chromatographers. Plots of residuals and moment
   analysis plots are particularly useful. Plotting of
   the same variable on both axes of the graph should be
   avoided.

o Error propagation analysis enables the error in a
   result to be determined based upon the estimated
   error in the components. It is often easy to apply and
   is revolutionizing SEC mathematical procedures. In
   particular, it shows that ratios of experimental data
   should generally be avoided. Also, it enables
   calculation of error in data to be fit by regression
   methods and aids in the assessment of the adequacy of
   the fitting procedure.

<u>Literature Cited</u>

1. Balke, S.T. "Quantitative Column Liquid Chromatography, A
    Survey of Chemometric Methods"; Elsevier: Amsterdam, 1984.
2. Delaney, M.F. <u>Anal. Chem.</u> 1984; 56, 261R.
3. Draper, N.R.; Smith, H. "Applied Regression Analysis"; John
    Wiley and Sons: New York, 1981, 2nd ed..
4. Parker, L.R., Jr.; Cave, M.R.; Barnes, R.M. <u>Analytica
    Chimica Acta</u> 1985, 175, 231.
5. Bandler, J.W. <u>IEEE Trans. on Microwave Theory Techn.</u> 1969,
    MTT-17, 533.
6. de Galan, L. <u>TRAC</u>, 1985, 4, 62.
7. Balke, S.T.; Hamielec, A.E.; LeClair, B.P.; Pearce, S.L.
    <u>Ind. Eng. Chem., Prod. Res., Dev.</u>, 1969, 8, 54.
8. Dawkins, J.V. In "Steric Exclusion Liquid Chromatography of
    Polymers"; Janca, J., Eds.; Marcel Dekker, Inc.: New York,
    1984; pp. 53-116.
9. Kubin, M. <u>J. Appl. Polym. Sci.</u>, 1982, 27, 2933.
10. Provder, T.; Rosen, E.M. <u>Separ. Sci.</u>, 1970, 5, 437.
11. Rosen, E.M.; Provder, T. <u>Separ. Sci.</u>, 1970, 5, 485.
12. Cardenas, J.N.; O'Driscoll, K.F. <u>J. Polym. Sci.</u>, 1975, B13,
    657.
13. Yau, W.W.; Malone, C.P. <u>J. Polym. Sci.</u>, 1967, B5, 663.
14. Scott, R.P.W. "Liquid Chromatography Detectors"; Elsevier:
    Amsterdam, 1977.
15. Carr, P.W. <u>Anal. Chem.</u>, 1980, 52, 1746.
16. Balke, S.T.; Hamielec, A.E. <u>J. Appl. Polym. Sci.</u>, 1973, 17,
    905.
17. Boni, K.A. In "Characterization of Coatings: Physical
    Techniques, Part II"; Myers, R.; Long, S.S., Eds.; Marcel
    Dekker: New York, 1976, pp. 68-121.

18. Rowe, P.N. Chemtech., 1974, January, 9.
19. Garcia-Rubio, L.H.; Talatinian, A.V.; MacGregor, J.F. Proc.
    Symposium on Quantitative Characterization of Plastics and
    Rubber; Vlachopoulos, J.; Ed.; McMaster University: Hamilton,
    Ontario, June 21-22, 1984, p. 37.
20. Ku, H.H. J. Res. Nat. Bur. of Stand., Sect. C, 1966, 70,
    263.
21. Bevington, P.R. "Data Reduction and Error Analysis for the
    Physical Sciences", McGraw Hill: New York, 1969.
22. Garcia-Rubio, L.H. J. Appl. Polym. Sci., 1982, 27, 2043.
23. Haefelfinger, P. J. Chromatogr., 1981, 218, 73.
24. Cheng, H.; Gadde, R.R. J. Chromatogr. Sci., 1985, 23, 227.

Appendix I: Error in Ratios of Experimental Quantities

Consider a result K, obtained by ratioing two experimental
quantities, A and c.

$$K = \frac{A}{c} \tag{I-1}$$

In the example analyzed by Garcia-Rubio (22), A was absorbance,
c, concentration, and K the extinction coefficient. The error in A
is random and quantified by the error variance, s_A^2. Similarly, the
error in c is given by s_c^2. Covariance is zero. That is, the
error of A can be determined independent of the error in c and vice
versa.

Application of the Error Propagation Equation (Equation 11)
gives:

$$s_K^2 = \left[\frac{1}{c}\right]^2 s_A^2 + \left[\frac{-A}{c^2}\right]^2 s_c^2 \tag{I-2}$$

In the best of circumstances, the error in A is negligible, so
that $s_A^2 = 0$ and:

$$s_K^2 = \frac{A^2}{c^4} s_c^2 \tag{I-3}$$

Writing this result as error bounds on K

$$K \pm 95\% \text{ Confidence Limits} = K \pm 2\, s_K \tag{I-4}$$

$$= \frac{A}{c}\left[1 \pm \frac{2s_c}{c} \right] \tag{I-5}$$

At low values of c the error boundaries explode in value,
regardless of high precision (low s_c) in c.

Appendix II: Calibration Curve Error

For a calibration curve of the form of Equation 1 the logarithm of the peak molecular weight of the standard is used and not simply the peak molecular weight. It is important to note that the error in the logarithm of the value is not the same as the error in the value itself.

This can be shown by application of the error propagation equation (Equation 11). If the error in the peak molecular weight M is represented by the error variance, s_M^2 and we let

$$y = \log M \qquad (II-1)$$

then the error in y is obtained by applying Equation 11 to Equation II-1 to obtain:

$$s_y = \frac{1}{2.303} \frac{s_M}{M} \qquad (II-2)$$

Thus, the error in the ordinate of the calibration curve, y, is actually a value proportional to the error in M divided by M. In other words, the fractional error in M is involved rather than the absolute error.

In using linear regression it is assumed that the error in the ordinate is a constant. Thus, we are justified in using linear regression to fit Equation 4 if the % error in M is constant but not if the absolute error in M is constant (because then the value of s_y will vary with M).

RECEIVED March 25, 1987

Chapter 13

Multiple Detectors in Size Exclusion Chromatography: Signal Analysis

L. H. Garcia-Rubio

College of Engineering, Department of Chemical Engineering, University of South Florida, Tampa, FL 33620

The characterization of complex polymers using size exclusion chromatography often requires the use of multiple detectors for the measurement of the polymer concentration and for the estimation of molecular properties such as composition and molecular weight. Depending upon the complexity of the analytical problem, a variety of detectors are used. Typical detection systems include refractometers, infrared and ultraviolet spectrophotometers, densitometers, on-line viscometers and low angle light scattering photometers. All of these detectors have different response factors and their measurement equations propagate the experimental error differently. Because of these differences, direct use of the synchronized signals from a system of detectors often leads to errors in the measurements and/or to partial interpretation of the data. This paper reports on the effect that the differences in detector sensitivity and measurement errors have on the quantitative interpretation of the data from multiple detector systems. As a case study, the signals from a detection system composed of a differential refractometer, a spectrophotometer and viscosity measurements on collected fractions are presented and discussed.

Complex polymers are those having a joint distribution of molecular properties. Branched polymers, copolymers and stereoregular polymers fall within this category. For example, branched polymers have a joint

0097-6156/87/0352-0220$06.00/0
© 1987 American Chemical Society

distribution of molecular weights and branching frequencies, copolymers have a joint distribution of composition/molecular weight/sequence length and tacticity. Complex polymers may result from homopolymerizations and copolymerizations. Thus, polybutadienes are complex polymers formed from the cis, trans and vinyl additions of butadiene. In addition, since polybutadienes are known to undergo branching reactions, polybutadienes are characterized by a joint composition/molecular weight/branching frequency distribution. In general, complex polymers do not exhibit a unique relationship between the size of the molecules in solution and the molecular weight as most linear homopolymers do. Therefore, the characterization of complex polymers using size exclusion chromatography (SEC) often requires the use of multiple detectors for the measurement of the polymer concentration and for the estimation of molecular properties such as composition, molecular weight and branching frequency (1-15). Depending upon the complexity of the analytical problem, a variety of detectors are used. Typical detection systems include refractometers (which have become the standard detectors in commercial instruments), infrared (2-4) and ultraviolet spectrophotometers (5-16), densitometers (17-19), on-line viscometers (7,10,20,21) and low angle light scattering photometers (22-25). All of these detectors have different response factors and their measurement equations propagate the experimental errors differently (26-28). Because of differences in sensitivity and noise level, direct use of the synchronized signals from a system of detectors often leads to errors in the measurements and/or to partial interpretation of the data (11,14,16). This paper reports on the effect that the differences in detector sensitivity and measurement errors have on the quantitative interpretation of the data from multiple detector systems. As a case study, the signals from a detection system composed of a differential refractometer, a spectrophotometer and viscosity measurements on collected fractions are presented and discussed.

Problem Formulation and Analysis

     The characterization of complex polymers requires at least one detector per desired property. Obviously, some of the molecular properties will be inter-related and the response from any given detector will include, in general, contributions from one or more properties. Thus, the response from ultraviolet and infrared spectrophotometers are known to contain information on the composition and the microstructure of the polymer chains (3,4,14,15,28,29), the intrinsic viscosity and light scattering measurements, on the other hand, respond to at least the composition and the molecular weight (20,24,25). The detailed discussion on the interpretation of complex signals has been presented elsewhere (26-29) and it will be discussed in the context of size exclusion chromatography in a separate publication. In what follows it will be assumed that each detector is trained to either a unique property or a linear combination of polymer properties.
     The minimum set of properties to be monitored in a chromatographic effluent are the polymer concentration, the polymer

composition and the molecular weight. The composition and polymer concentration detectors provide information that is, in general, indepenednt of the molecular weights. The interpretation of the signals from molecular weight detectors (ie: LLALS and on-line viscometers) requires prior knowledge of the polymer concentration. A significant fraction of the error in the interpretation of molecular weight detectors is largely the result of propagated measurement errors arising from the estimation of both the polymer composition and the polymer concentration (14-16,27). Figure 1 shows the effect of the difference in sensitivity for the concentration and composition detectors. As it can be seen in the regions of low signal to noise ratio (i: the tails of the chromatogram), there is an apparently dramatic increase in the content of one of the comonomers, styrene in this case. Clearly, it is important to assess how much of the signal represents actual changes in composition and how much it is due to the differences in sensitivity and noise level of the detectors involved. The types of deviations shown in Figure 1 are typical of measurements conducted with detectors having different sensitivities (see references cited in 1 and 14-16). The deviations observed have been attributed to the effect of the microstructure on the detector response, however, no attempt has been made to estimate the error bounds of the measurements and thus elucidate the extent to which the deviations observed can be attributed to microstructure or compositional effects on the polymer fractionation process and on the detector responses.

If a chromatograph equipped with a refractometer, a spectrophotometer, a viscometer and a LLALS photometer is assumed for the analysis of a two component polymer system, the equations relating the composition and the polymer concentration to the refractometer and the spectrophotomer are given by (1,14-16)

$$n = \left[\nu_1 P_1 + \nu_2(1 - P_1)\right] C \qquad (1)$$

$$A = \varepsilon_2(1 - P_1) C \qquad (2)$$

Where n is the refractive index difference between the polymer in solution and the solvent, ν_i is the specific refractive index increment or refractometer response to the ith species, P_i is the weight fraction of the ith species, A is the absorbance, ε_i is the absorption coefficient or the spectrophotometer response to the ith species and C is the total concentration of polymer in the effluent stream. In this case, it has been assumed that only one of the components absorbs at the frequency selected for the spectrophotometer.

The equations relating signals from the molecular weight detectors to the concentration and the molecular weights are given by (22)

$$KC/R(\theta) = 1/Mw + 2A_2 C \qquad (3)$$

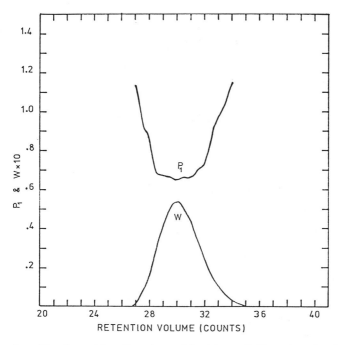

Fig. 1. Results from the direct application of the signals from a SEC detection system composed of a differential refractometer and uv spectrophotometer (254 nm).

Where Mw is the weight average molecular weight or the apparent molecular weight in the case of copolymers, A_2 is the second virial coefficient, and K is an optical constant defined as

$$K = \frac{2\pi^2}{N} \left(\frac{n_0}{\lambda_0^2}\right)^2 \left(\frac{dn}{dC}\right)^2 (1 + \cos^2 \theta) \qquad (4)$$

$R(\theta)$ is the excess Rayleigh ratio, n_0 and λ_0 are the refractive index of the solvent and the wavelength of the light source.

If the second virial coefficient is negligible, equation 3 becomes

$$R(\theta)/K = MwC \qquad (5)$$

In the case of intrinsic viscosity measurements, the measurements will be in the form of pressure drop or a ratio of efflux times. In either case, the interpretation equation is of the form

$$Q = [\eta]C \qquad (6)$$

Where $Q = \ln(\Delta\Pi_1/\Delta\Pi_0)$ when pressure transducers are used (20) and it will be defined as $Q = (t - t_0)/t_0$ when capillary viscometers are used (7).

The desired polymer properties can be obtained directly from equations 1,2,5 and 6

$$P_1 = [\varepsilon_2 - \nu_2(A/n)]/[\varepsilon_2 - (\nu_2 - \nu_1)(A/n)] \qquad (7)$$

$$C = [\varepsilon_2 n - (\nu_2 - \nu_1)A]/(\varepsilon_2 \nu_1) \qquad (8)$$

$$Mw = R(\theta)/(KC) \qquad (9)$$

$$[\eta] = Q/C \qquad (10)$$

Note that all the desired properties depend directly or indirectly on the ratio A/n and therefore, on the relative response factors and signal to noise ratios of the spectrophotometer and the differential refractometer. In general, spectrophotometers are more sensitive than differential refractometers, therefore, the ratio $A/n \rightarrow \infty$ at the tails of the chromatogram, that is, the refractometer signal will be zero while there still be a signal from the spectrophotometer. In addition, equations 9 and 10 are hyperbolic functions of the concentration. Thus, as the concentration decreases the apparent values of Mw and $[\eta]$ will increase, increasing the uncertainty in the estimates of the molecular weights and the intrinsic viscosities (see Appendix I and references 26-29). In the limit when $A/n \rightarrow \infty$ (or $n/A \rightarrow 0$), the polymer composition and the total polymer concentration have finite values that are different from the limits expected on the basis of the mass balances

$$\lim_{(A/n)\to\infty} P_1 = 1/(1 - \nu_1/\nu_2) \qquad (11)$$

$$\lim_{(A/n)\to\infty} C = -(\nu_2 - \nu_1)A/(\epsilon_2\nu_1) \tag{12}$$

That is, depending on the magnitude and sign of the response factors, the weight fraction can be greater than one or it can be negative. For example, if $\nu_1 < \nu_2$, in the limit where the refractometer signal has approached zero, the fraction of monomer 1 in the copolymer will be greater than 1. The same applies to the concentration, it can be greater than the concentration injected in the chromatograph or it can also be negative depending on the values of the response factors. It is important to realize that the effects suggested by equations 11 and 12 will appear as deterministic trends in the data, thus misleading the estimation of the molecular weights and the interpretation of the composition data (26,27). It is clear that the limiting factor in the interpretation of the signals from a multidetector system will be the detector with the least sensitivity. When this signal becomes zero or it is confounded with the instrument noise, the system of equations 7-10 becomes under-determined. In addition to the problems associated with the differences in detector sensitivity, the measurement noise also plays an important role. As it can be seen in Figure 2, the instrument noise propagated through equations 7-10 can cause wild oscillations in the regions of low signal-to noise ratio (ie: the tails of the chromatograms). Clearly, without good estimates of the polymer concentration and the polymer composition, the estimates of both the point values and the average molecular properties will be considerably biased. In some cases it is possible to solve the problem experimentally by collecting fractions and analyzing them independently for concentration and composition (14-16), unfortunately, this is not a satisfactory solution. It is always possible, however, to obtain good estimates of the errors associated with the measured polymer properties (ie: composition, molecular weights, etc.) The application of simple error propagation techniques (26,27,30) to equations 7-10 yields good approximations to the variance of the desired properties

$$\text{var}(P_1) = \left(\frac{\epsilon_2\nu_1}{n}\right)^2 \left[\frac{(A/n)^2 \text{var}(n) + \text{var}(A)}{(\epsilon_2 - (\nu_2 - \nu_1)(A/n))^4} \right] \tag{13}$$

$$\text{var}(C) = \frac{\epsilon_2{}^2 \text{var}(n) + (\nu_2 - \nu_1)^2 \text{var}(A)}{(\epsilon_2\nu_1)^2} \tag{14}$$

$$\text{var}(Mw) = \left(\frac{1}{KC}\right)^2 \text{var}(R(\theta)) + \left(\frac{R(\theta)}{KC^2}\right)^2 \text{var}(C) \tag{15}$$

$$\text{var}(\eta) = \left(\frac{1}{C}\right)^2 \text{var}(Q) + \left(\frac{Q}{C^2}\right)^2 \text{var}(C) \tag{16}$$

From equations 13-16, the standard error for each measurement as a function of the elution time can be obtained. Additional propagation of these errors through the integration across the chromatogram results in estimates of the errors associated with the SEC calculation of the average polymer properties. Therefore, it enables reliable statistical comparisons between SEC estimates and static measurements

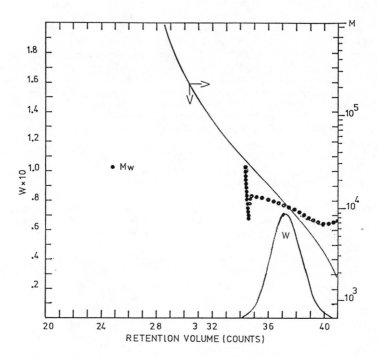

Fig. 2. Effect of the uv detector noise on the estimation of the molecular weight as function of elution volume for a narrow polystyrene standard. The symbols represent molecular weights obtained using the universal calibration and viscosity measurements on collected fractions

on the total polymer. For example, if the Universal Calibration is used and the chromatogram is integrated using the trapezoidal rule,

$$Mw = \sum_{1}^{n} M(i)[w(i) + w(i+1)] \, \Delta v/2 \qquad (17)$$

$$M(i) = J(i)/[\eta]i \qquad (18)$$

where Δv represents the elution volume increment between sample points, $w(i)$ is the weight fraction of polymer at the ith sample point, $J(i)$ is the polystyrene-based universal calibration curve and $[\eta]i$ is the intrinsic viscosity at the ith sample point. The variance for the weight average molecular weight can be approximated as

$$var(Mw) = \left(\frac{\Delta v^2}{4}\right) \sum_{1}^{n}\left[M(i) \, var(w(i)) + \right.$$

$$\left. (w(i) + w(i+1)) \left(\frac{J(i)}{[\eta]i^2}\right)^2 \, var([\eta]i)\right] \qquad (19)$$

The variance for the weight fraction $w(i)$ can be obtained from equation 14 and the variance for $M(i)$ can be calculated directly from equation 18. Note that the propagation of error analysis can be readily extended to other averages and and it can also be used to account for the errors associated with the calibration of columns and detectors.

Experimental Methods

Styrene-acrylonitrile copolymers were synthesized by bulk free-radical polymerization at 60°C and 40°C using AIBN as initiator. The Copolymer compositions were determined by gas chromatography and verified by ^{1}H NMR spectroscopy. The copolymers were purified prior to the SEC analysis by precipitation from absolute methanol. The size exclusion chromatograph is a modified room temperature Waters high pressure liquid chromatograph equipped with a differential refractometer and a Waters 440 dual uv detector. The injection valve, detectors and volume counter were interphased with a Data General Nova mini computer that was used primarily for data logging, base line corrections, and data synchronization. The first detector downstream from the columns was used as reference for synchronization purposes. The refractometer and uv spectrophotometers were calibrated to absolute units using benzene-CCl$_4$ solutions. Fractions were collected at regular intervals for intrinsic viscosity measurements. Intrinsic viscosities were measured with Cannon Ubbelohde viscometers at 25°C. The SEC was equipped with a set of six μ-Styragel columns (100, 500, 10^3, 10^4, 10^5, and 10^6 Å). [μ-Styragel is a registered trademark of Waters Associates.] For size exclusion chromatography analysis a 200 μL sample volume of 0.5% (w/v) concentration was injected onto the columns. The solvent flow rate was set at 1ml/min. THF was used as mobile phase. The columns were calibrated using 18 narrow polystyrene standards [Pressure Chemicals]. The molecular weights were calculated using the universal calibration curve principle and reported Mark-Houwink parameters. Whenever fractions were collected the molecular weights were calculated directly from the universal calibration and intrinsic viscosity measurements.

Results and Discussion

Polymers produced at various initial monomer compositions and conversion levels were analysed using equations 7,8 and 10. The 95% confidence interval (2σ) calculated from the corresponding error propagation equations has been superimposed on to the measured property values. The results are shown in figures 3-5. Figure 3 shows a low conversion copolymer synthesized with a feed composition containing 60% styrene (ie:f_{10} = 0.6). It is expected, from the theory of copolymerization, that this polymer would have a uniform composition across the chromatogram. The direct interpretation of the synchronized signals suggests that this is not the case. As it can be seen at the tails of the chromatogram (ie: at the low and high molecular weight ends) there are significant changes in composition suggesting that the Mayo-Lewis kinetics are not valid. However, application of the error propagation equations indicates that, given the variance of the measurements, there is no evidence to suggest that the compositions at the tails of the chromatogram are different from the values at the middle of the chromatogram, which agree with the expected composition from copolymerization kinetics. Figure 4 shows the data for a high conversion high styrene content copolymer (ie: f_{10}=0.9, P_1 = 0.85). In this case it appears that there is evidence of the intermediate size chains having statistically significant differences in composition. Compositionally heterogeneous polymer chains can be expected at high conversions since high conversion copolymers may not follow the copolymerization theory due to the gel effect. Finally, Figure 5 shows a low conversion low styrene copolymer (f_{10}=0.4, P_1 = 0.55) for which the molecular weights were determined from intrinsic viscosity measurements on collected fractions. Clearly, the calculated error bars (ie: the 95% confidence interval) at each sampling point allows a realistic interpretation of the SEC data. The calculation of the average Mn, Mw and composition and their corresponding 95% confidence interval for the whole polymer allows the statistical comparison between the SEC estimated averages and the static measurements on the whole polymer. The error analysis on the static measurements was done in accordance with the equations in Appendix I and reference 27. As it can be seen in Table I, if allowance is made for the propagated experimental errors, there is good agreement between the SEC and the static measurements at the 95% confidence level.

Summary and Conclusions

The direct interpretation of the signals from several detectors in SEC experiments has been analysed. It is clear that the direct interpretation of the synchronized signals from detectors having

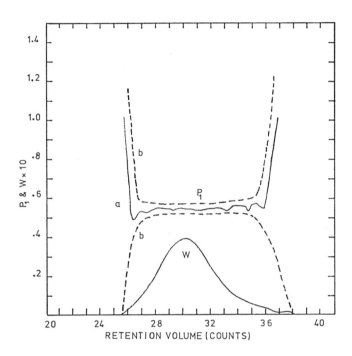

Fig. 3. Interpretation of the signals from a multidetector SEC: a. direct calculations; b. 95% confidence interval included. Low conversion, intermediate styrene content SAN copolymer.

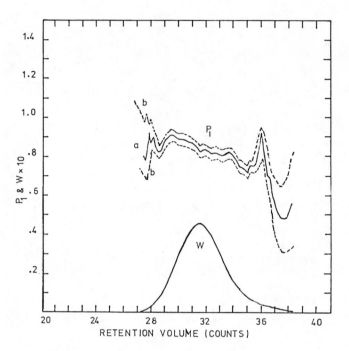

Fig. 4. Interpretation of the signals from a multidetector SEC: a. direct calculations; b. 95% confidence interval included. High conversion styrene-rich SAN copolymer.

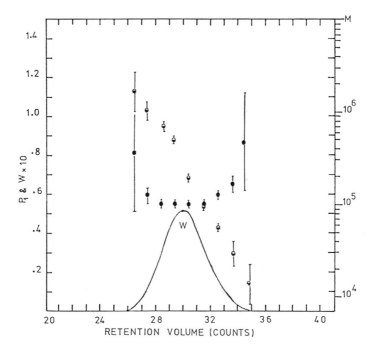

Fig. 5. Interpretation of the signals from a multidetector SEC: direct calculations;   95% confidence interval included. Low conversion, low styrene content SAN copolymer: ● styrene content,◐ molecular weight.

TABLE I

COMPARISON BETWEEN STATIC MEASUREMENTS AND SEC ESTIMATES

Sample	Static Measurements			SEC Results	
	P_1 GC	[n](dl/g) THF 25°C	Mw Viscosity	Mw	[n]
standard deviation	0.01	0.025	2.0%	3.0%	0.05
R0209	0.87	0.406	80031	81190	0.500
R0204	0.84	0.533	92955	99061	0.510
R0206	0.80	0.588	120151	109240	0.725
R0208	0.74	0.790	124305	114680	0.874
R0210	0.69	0.963	127264	136300	0.948
R0211	0.68	1.048	141323	153060	1.053

different sensitivities and noise levels can lead to ambiguities in the interpretation of SEC data. The application of simple error propagation techniques results in equations for the estimation of the variances of the desired polymer properties. The propagated experimental errors can then be used to eliminate ambiguities in the data and for the statistical comparison of static measurements on the whole polymer and the estimates from SEC experiments. The error propagation analysis can be readily applied to more complicated measurement equations (ie: equation 3) provided that there are sufficient measurements to estimate all the parameters involved. Furthermore, if detectors like the diode array spectrophotometers are used, several measurements on the same variable will be obtained, thus allowing for the statistical estimation of the desired polymer properties.

Acknowledgments

This research was supported by the NSF Initiation Grant RII-8507956 The author is indebted to Dr. T. Provder from SCM Corporation for his continuous encouragement and support.

REFERENCES:

1. Yau, W.; Kirkland, J.; Bly, D., "Modern Size-Exclusion Liquid Chromatography", 1979, J. Wiley, New York.
2. Terry, S. L.; Rodriguez, F. J. Polym. Sci., Part C 1968, 21, 103.

3. Mirabella, T.; Barral, E.; Johnson, J. J. Appl. Polym. Sci. 1975, 19.
4. Mirabella, T.; Barral, E.; Johnson, J. J. Appl. Polym. Sci., 1976, 20.
5. Runyon, J. R. et al, J. Appl. Polym. Sci.,1969, 13, 2359.
6. Adams, H. E. In "Gel Permeation Chromatography"; Altgelt, K.; Segal, L., Eds.; Dekker: New York, 1971.
7. Grubisic-Gallot, Z. et al. J. Appl. Polym. Sci. 1972, 16, 2933.
8. Probst, J. J.; Unger, K.; Cantow, H. J., Angew. Makromol. Chem., 1974, 35, 177.
9. Elgert, K. F.; Wohlschiess, R., Angew. Makromol. Chem., 1977, 57, 87.
10. Teramachi, S.; Hasegawa, A.; Yoshida, S., Macromolecules., 1983, 16, 542.
11. Balke, S. T.; Patel, R. D. In "Size Exclusion Chromatography (GPC)" Provder, Theodore, Ed.; ACS SYMPOSIUM SERIES No. 138; ACS; Washington, D.C., 1980; pp. 149-82.
12. Inagaki, H.; Tanaka, T., Pure and Appl. Chem., 1982, 54(2), 302.
13. Teramachi, S. Hasegawa, A.; Yoshida, S., Macromolecules., 1983, 16, 842.
14. Mori, S.; Suzuki, T., J. Liq. Chrom., 1981, 4(10), 1685.
15. Garcia-Rubio, L. H.; MacGregor, J. F.; Hamielec, A. E., ACS SYMPOSIUM SERIES, T. Provder, Ed., 1982, 197, 151.
16. Garcia-Rubio, L. H.; Hamielec, A. E.; MacGregor, J. F., ACS SYMPOSIUM SERIES, C. D. Craver Ed., 1983, 202, 310.
17. Francois, J.; Jacob, M.; Grubisic-Gallot, Z.; Benoit, H., J. Appl. Polym. Sci., 1978, 22, 1159.
18. Trathing, B.; Jorde, C., J. Chromatogr., 1982, 24, 147.
19. Boyd, D.; Narasimhan, V.; Huang, R.; Burns, C.M., J. Appl. Polym. Sci., 1984, 29, 595.
20. Ouano, A. C., J. Polym. Sci: Symp. Series., 1973, 43, 299.
21. Lecheaux, D. et al., J Appl. Polym. Sci., 1982, 27, 4867.
22. Ouano, A. C.; Kay, W., J. Poly. Sci., Part A-1, 1974, 12, 1151.
23. Hamielec A. E., J. Liq. Chrom., 1980, 3, (3), 381.
24. MacRury, T. B.; McConell, M. L., J. Appl. Polym. Sci., 1979, 24, 651.
25. Malihi, F. B.; Kuo, C. Y.; Provder, T., J. Appl. Polym. Sci., 1984, 29, 925.
26. Garcia-Rubio, L. H., J. Appl. Polym. Sci., 1982, 27, 204.
27. Garcia-Rubio, L. H.; Talatinian, A. V.; MacGregor, J. F., Proc. Symp. on Quantitative Characterization of Plastics and Rubber, 1984, McMaster University, 37-55.
28. Garcia-Rubio, L. H.; Ro, N.; Patel, R., Macromolecules, 1984, 17, 1998.
29. Garcia-Rubio, L. H.; Ro, N., Can. J. Chem, 1985, 63(1), 253.
30. Himmelblau, D. M., "Process Analysis by Statistical Methods"., 1970, John Wiley and Sons, New York.
31. Sutton, P. M.; MacGregor, J. F., Can. J. Chem. Eng., 1977, 55, 602.
32. Reilly, P. M.; Patino-Leal, H., Technometrics, 1981, 23(3), 221.

APPENDIX I

ERRORS IN THE ESTIMATION OF POLYMER PROPERTIES

In this appendix, a summary of the error propagation equations and objective functions used for standard characterization techniques are presented. These equations are important for the evaluation of the errors associated with static measurements on the whole polymers and for the subsequent statistical comparison with the SEC estimates (see references 26 and 27 for a more detailed discussion of the equations). Among the models most widely used to correlate measured variables and polymer properties is the truncated power series model

$$y = \sum_{i=1}^{N} k_i x^i \tag{I1}$$

where y is the measured quantity (efflux time, intensity of light etc.) and x is the independent variable, normally concentration. The parameters k_i are generally interpreted in terms of one or several molecular properties. The first-order effects (k_i) represent "specific polymer properties" such as, intrinsic viscosities, specific refractive index increments, first virial coefficients etc. which are related directly or indirectly to primary molecular properties like the number and weight, average molecular weights, chemical composition etc. Second and higher-order effects (ie: i>1) are related to intermolecular effects such as hydrogen bonding and end-to-end molecular distances. The second-order effects are often used to explain the behaviour of polymer solutions and polymer melts. The problems associated with the estimation of polymer properties using equation I1 can be divided into two closely-related statistical problems: the propagation of experimental errors through the measurements and measurement equations and the problem of model discrimination and parameter estimation .
In order to analyze the results from polymer characterization measurements, equation I1 is, normally, transformed into its "specific form" i.e.:

$$z = \frac{y}{x} = \sum_{i=1}^{N} k_i x^{i-1} \tag{I2}$$

The rationale behind this transformation has been that, for most cases, the first two terms in the series are sufficient to adequately represent the measurement as a function of the independent variable i.e.:

$$z = \frac{y}{x} = k_1 + k_2 x \tag{I3}$$

Note that, the transformation y/x implies an error transformation. Therefore, the variable z may show trends which are due to the experimental error. In addition, plots of z (y/x) vs x have implicit correlations since z will vary with x. This correlation may also appear in the form of trends, which can be particularly misleading in the assessment of the number of terms required by equation I-1 to represent the data.

If σy^2 is the variance of the measurement y and σx^2 is the variance of the independent variable x, then for the case of independent measurements in x and y, the variance of the correlation variables y or z can be approximated by

$$\text{var} (y) = \sigma y^2 \tag{I4}$$

$$\text{var} (z) = \sigma y^2 \; \frac{1}{x}^2 + \sigma x^2 \; \frac{y}{x^2}^2 \tag{I5}$$

from equations I4 and I5, it can be shown that:

$$\text{var} (z) > \text{var} (y) \quad \text{for all x such that} \quad 0 < |x| \leq 1$$

$$\text{var} (z) = \text{var} (y) \quad \text{for all x such that} \quad x = x_0 \text{ and}$$

$$\frac{\sigma y^2}{y^2} = \frac{\sigma x^2}{x_0^4 - x_0^2} \tag{I6}$$

$$\text{var} (z) < \text{var} (y) \quad \text{for all x such that} \quad |x| > |x_0|$$

Equations I4 through I6 indicate that careful consideration must be given to the analysis of the error structure and to the confidence intervals of the correlation variables, otherwise trends due to errors will appear as real when, in fact, they may be contained within the error bands of either y or z.

Sources of Experimental Error

There are two main sources of error propagation in static measurements, errors due to successive dilutions and errors due to initial instrument offset. Other errors which are also applicable to SEC analysis are discussed in (1). These errors can be propagated using the criteria presented here. If w is the intial mass of polymer and V_1 is the amount of solvent added to obtain the desired concentration Ci, the dilution process can be represented by the following set of equations:

$$C_1 = \frac{w}{V_1} \tag{I7}$$

$$C_2 = \frac{w}{V_1 + V_2} \tag{I8}$$

$$\ldots \; C_n = \frac{w}{\displaystyle\sum_{j=1}^{n} V_j}$$

In the same dilution is used n times, (i.e. the same pipette with constant % error ε is used).

$$C_n = \frac{w}{V_1 . n(1+\varepsilon)} = \frac{C_1}{n(1 + \varepsilon)} \tag{I9}$$

where n is now the dilution factor. The error associated with the dilution process is a function of the initial error in w and the cumulative error due to volumetric additions. From equation I9 it follows directly that the variance and the covariance of C_n can be approximated by

$$\text{var} \; (C_n) = \frac{\text{var}(w)}{n^2 V_1^2} + \left(\frac{w}{n^2 V_1^2}\right)^2 n \; \text{var}(V_1) \tag{I10}$$

$$\text{cov} \; (C_i, C_j) = \left(\frac{1}{iV_1}\right)\left(\frac{1}{jV_2}\right) \text{var}(w) + \left(\frac{C_i^2}{w}\right)\left(\frac{C_j^2}{w}\right) i \; \text{var}(V_1) \tag{I11}$$

It can be seen form equations I9 to I11 that both the variance and the covariance terms decrease with increasing dilution (n) keeping approximately a constant error for small dilution errors (ε). A source of error seldom considered is the initial offset (ε_0). This initial bias may be due to an instrument offset, a change in the reference values or to an error in the initial concentration and it causes a constant displacement along the x or y axis, i.e.:

$$y = k_1 x + k_2 x^2 + \varepsilon_0 \tag{I12}$$

or for the specific model:

$$z = \frac{y-\varepsilon_0}{x} = k_1 + k_2 x \tag{I13}$$

The initial offset can give raise to deterministic trends whenever successive dilutions are employed (27).

Choice of Objective Functions

Since both x and y contain experimental error, the parameter estimation problem should be treated as an error in variables problem (3,32). The objective function to be minimized is given by:

$$\Theta = \sum_{j=1}^{M} \left(\frac{\varepsilon_j}{\text{var}(\varepsilon_{jj})} \right)^2 \tag{II4}$$

where

$$\varepsilon_j = y_j - \sum_{i=1}^{n} k_i x_j^i + \varepsilon_0$$

and

$$\text{var}(\varepsilon_j) = \sigma y^2 = \frac{\partial \varepsilon_j}{\partial x_j}^2 \sigma x^2$$

Similar objective functions can be constructed for the "specific" model (equation I2).

APPENDIX II

CALIBRATION OF THE SEC DETECTION SYSTEM

The main objectives in calibrating the SEC detection system in absolute refractive index and absorption units are the estimation of ν and ε at the normal flow conditions and the standardization of the measurement errors. The first step in the calibration process is the estimation of the instrument's constants to transform the computer units into absorbances and refractive index units. The Waters 440 UV spectrophotometer displays absorbance units. Therefore, step changes in the instrument's balance and sampling of the signal provide the necessary data for the calibration. The equations obtained are:

$$\Delta A_1 = (0.118685 \times 10^{-3}) \cdot cu_1 \tag{II1}$$

$$\Delta A_2 = (0.122639 \times 10^{-3}) \cdot cu_2 \tag{II2}$$

where $\Delta A(i)$ is the absorption difference between the sample and the base line for the ith detector and $cu(i)$ is the absolute difference in computer units (409.5 cu = 1 volt) between the base line and the signal for the ith detector. The calibration of the differential refractometer requires solutions of known refractive index difference. Benzene/CCl_4 solutions are known to follow the ideal behaviour (volumes are additive). In addition, the refractive index of these mixtures is proportional to the volume fraction of benzene, therefore:

$$n_{sol} = n_B x_B + (1-x_B) n_{CCl_4} \tag{II4}$$

$$\Delta n = (n_{sol} - n_{CCl_4}) = (n_B - n_{CCl_4}) * x_B$$

n_{sol} = refractive index of the solution

n_B = refractive index of benzene (1.5020)

n_{CCl_4} = refractive index of CCl_4 (1.4596)

x_B = volume fraction of benzene

Using step changes in the concentration of benzene, the response was correlated with the calculated refractive index difference (ΔRI) at an attenuation of 4x.

$$\Delta RI = (2.537534 \times 10^{-8}) .cu \qquad (II5)$$

Using benzene as a tracer, it is possible to estimate the volume of the injection loop under the assumption that all the material injected leaves the columns in the interval (to – tf). The mass of material injected can be calculated by integrating the contents of the detector cell. A simple mass balance gives

$$M_o = C_i\ V_{loop} = \int_{to}^{tf} \frac{\Delta n(t)}{v} . v(t)\ dt \qquad (II6)$$

If an average value of flow rate is used (v) equation II6 becomes

$$M_o = C_i\ V_{loop} = \frac{v}{\nu} \int_{to}^{tf} \Delta n(t)\ dt \qquad (II7)$$

where

C_i = concentration of the injected solution

$\nu = (n_B - n_{CCl_4})$

M_o = mass injected

The volume V_{loop} was estimated using equation II7 and replicated injections of various concentrations is

$$V_{loop} = 2.22 \pm 0.12\ cm^3.$$

Once the volume of the injection loop is known, the overall extinction coefficient (ε) and refractive index increment (ν) for an unknown sample can be calculated directly from equation II6 and the corresponding equations for the uv spectrophotometer at the required wavelengths

$$\nu = \frac{v}{C_i \, V_{loop}} \int_{to}^{tf} \Delta n(t) . dt \tag{II8}$$

$$\epsilon_\lambda = \frac{v}{C_i V_{loop}} \int_{to}^{tf} \Delta A(t) . dt \tag{II9}$$

The average flow rate was estimated from averaging the time elapsed between consecutive counter signals.

RECEIVED May 4, 1987

Chapter 14

Light Scattering Characterization of Branched Polyvinyl Acetate

Q.-W. Wang, I. H. Park, and B. Chu[1]

Department of Chemistry, State University of New York, Stony Brook, NY 11794—3400

Branching effects of polyvinyl acetate (PVAc) were investigated in a good solvent, methyl ethyl ketone at 25°C, by laser light scattering, high performance size exclusion chromatography (SEC) and viscometry. Several branched PVAc samples were synthesized and the degree of branching was characterized using the radius of gyration (R_g) and the effective hydrodynamic radius (R_h) by means of light scattering, as well as the hydrodynamic volume parameter by means of intrinsic viscosity ([η]). The branching effect on the size parameters (R_g and R_h) was much smaller than that on the hydrodynamic volume parameter. In particular, the change of the ratio R_g/R_h with respect to the degree of branching was negligibly small (maximum of a few percent even for the most highly branched PVAc). We were able to obtain the molecular weight distribution (MWD) of branched PVAc using the Laplace inversion of the intensity-intensity time autocorrelation function and static light scattering measurements. The results are in good agreement with the MWD obtained from a combination of SEC and intrinsic

Long-chain branched polymers, such as polyvinyl acetate (PVAc) and low density polyethylene (LDPE), are important technologically as well as scientifically in polymer rheology. (1,2) Different analytical methods, such as size exclusion chromatography (SEC)/viscometry, (3-5) SEC/light scattering and SEC/ultracentrifugation, (6,7) have been used to characterize the degree of long-chain branching. In particular, the SEC/viscometry method has been applied extensively to analyze the branching effects. More recently Chu, et al.(8) tried to estimate the branching effect and the molecular weight distribution (MWD) of branched LDPE using a combined technique of static light scattering (SLS) and dynamic light scattering (DLS). In this article, we want to examine the branching effects on the

---

[1]Correspondence should be addressed to this author.

0097—6156/87/0352—0240$06.50/0
© 1987 American Chemical Society

determination of MWD of branched PVAc using DLS and to check our analysis and results using the SEC/viscometry method.

Experimental

1. Synthesis of PVAc

Vinylacetate monomer of polymerization grade (over 99.98% purity) was used for the polymerization of branched PVAc. Samples B_4, B_3 and B_2 were polymerized at 60°C for 7 days, 4 days and 2 days, respectively. Sample B_1 was synthesized at room temperature for 2 months. Some branched PVAc samples were converted to linear polyvinyl alcohol (PVA) through saponification with a methanol solution of NaOH. The product, PVA, was then washed with methanol and reacetylized by anhydride acetic acid in pyridine (solvent) at ~95°C until all of the PVA solid dissolved and reacted. A trace of pyridine was removed from the linear PVAc by precipitating the linear PVAc several times using acetone and water. Each PVAc sample was fractionated into about 10 fractions using acetone as a good solvent and water as the non-solvent. B_{41} and B_{21} represent the highest molecular weight fractions among the branched PVAc prepared while B_{14} and B_{34} denoted the fourth fraction of B_1 and B_3 samples, respectively. It should be noted that free-radically synthesized PVAc is a complex mixture of linear and branched chains. The branched chains have different branching frequencies and branched lengths even after fractionation. Therefore, our fractions do not represent well-defined polymer fractions of uniform branching frequencies and lengths.

2. SEC Measurements

A set of two Waters ultrastyragel columns, designated 10^5 Å and 10^6 Å and a Waters pump (Model 590) for HPLC were used in this study. The elution solvent was tetrahydrofuran (THF) which was distilled in the presence of a small amount of CaH_2 in order to remove the peroxide. The flow rate was maintained at 1 ml/min. The sample injection volume was ~30 μl. The chromatogram detected by the differential refractometer (Waters R401) was recorded on a strip chart recorder. All experiments were performed at room temperatures with concentrations below the over-loading condition.

3. Light Scattering Measurements

We used an argon ion laser (Lexel Model 95) operating at λ_o = 514.5 nm. A standard photon counting detection system was used to measure the intensity of scattered light while the single-clipped photoelectron count auto-correlation function was measured with a Malvern Loglin 7027 digital correlator. All sample solutions for light scattering experiments were filtered through a Millipore Teflon filter of nominal 0.2 μm pore size. The filtrate was again centrifuged for ~2 hours at ~4000 gravity and finally transferred into the light scattering cell using a dust-free pipet. We used benzene as a standard for computing the Rayleigh ratio R_{vv} and took R_{vv} = 2.49x10^{-5} cm^{-1} for benzene at θ = 90°, λ_o = 514.5 nm and 25° C. The refractive index, the refractive index increment, the density, and the viscosity of methyl ethyl ketone (MEK) at 25°C were 1.3760 at λ_o = 514.5 nm, 0.081 ml/g at λ_o = 514.5 nm,(9) 0.80 g/ml and 0.401 cp, respectively.

4. Methods of Data Analysis

4.1 Light scattering data.

(a) Intensity of scattered light.

From measurements of scattered light intensity as a function of concentration and scattering angle, we can determine the weight average molecular weight, M_w, the z-average radius of gyration, R_g and the second virial coefficient, A_2. According to the classical Rayleigh-Debye theory, we have

$$\frac{HC}{R_{vv}(\theta)} = \frac{1}{M_w}\left(1 + \frac{16\pi^2 n_o^2}{3\lambda_o^2} R_g^2 \sin^2(\theta/2)\right) + 2A_2 C \qquad (1)$$

where $H = (4\pi^2 n_o^2/N_a \lambda_a^4)(\partial n/\partial C)^2$ with λ_o, n_o, N_a and $(\partial n/\partial C)$, being, respectively, the wavelength of light in vacuum, the solvent refractive index, Avogadro's number and the refractive index increment. $K[\equiv (4\pi/\lambda)\sin(\theta/2)$ with $\lambda = \lambda_o/n_a]$ is the magnitude of the momentum transfer vector. $R_{vv}(\theta)$ is the excess Rayleigh ratio of the solution at concentration C and scattering angle θ. The intercept at $\theta = 0$ and C=0 yields the weight average molecular weight, M_w. The slope in a $(HC/R_{vv})_{\theta\to 0}$ vs concentration plot and the angular dependence of $(HC/R_{vv})_{C\to 0}$ yield the second virial coefficient A_2 and the z-average radius of gyration, R_g, respectively.

(b) Spectrum of scattered light.

The spectrum of scattered light contains dynamical information related to translational and internal motions of polymer chains. In the self-beating mode, the intensity-intensity time correlation function can be expressed (10) as

$$G^{(2)}(t) = \langle I(0)I(t)\rangle = A(1 + b|g^{(1)}(t)|^2) \qquad (2)$$

where A is the base line, b is a coherence function of the spectrometer, $g^{(1)}(t)$ is the normalized time correlation function of the scattered electric field and t is the delay time. Although $g^{(1)}(t)$ is a single exponential decay curve for monodisperse, non-interacting and structureless macromolecules, $g^{(1)}(t)$ is related to the normalized characteristic linewidth distribution $G(\Gamma)$ through a Laplace integral equation for polydisperse systems by eq. (3).

$$g^{(1)}(t) = \frac{\langle E_s^*(0)E_s(t)\rangle}{\langle E_s^*(0)E_s(0)\rangle} = \int_0^\infty G(\Gamma)e^{-\Gamma t}d\Gamma \qquad (3)$$

where Γ is the characteristic linewidth. For narrow size distributions, the second-order cumulant expansion can be used to determine the average characteristic linewidth $\bar{\Gamma}$

$$\ln\left| g^{(1)}(t) \right| \simeq -\bar{\Gamma}t + 1/2\ \mu_2 t^2 \tag{4}$$

where the variance is $\mu_2/\bar{\Gamma}^2$ with $\mu_2 = \int (\Gamma - \bar{\Gamma})^2 G(\Gamma) d\Gamma$. Eq. (3) is valid for polydisperse particles with internal motions. Then $G(\Gamma)$ becomes a complex function of both size and internal motions which, under favorable conditions, may be separated by examining the angular dependence of $G(K,\Gamma)$. In our case, we know that as $K{\to}0$, internal motions become less important and measurements of $G(\Gamma)$ at low scattering angles reveal essentially only a distribution of translational diffusion coefficient ($D \equiv \Gamma/K^2$). The presence of internal motions and the interference effect can also be represented as a K^2 expansion for $\bar{\Gamma}$.(11) At a finite scattering angle θ and a finite concentration C, we have

$$\bar{\Gamma}/K^2 = \bar{D}_z^0\ (1 + f\ R_g^2 K^2)(1 + \bar{k}_d C) \tag{5}$$

where $\bar{D}_z^0$ is the z-average translational diffusion coefficient at infinite dilution, f is a constant depending upon the chain structure, polydispersity and solvent quality; and $\bar{k}_d$ is an average diffusion second virial coefficient. In order to estimate the normalized linewidth distribution $G(\Gamma)$, we chose an algorithm developed by Provencher(12)and commonly known as CONTIN. At $t=0$, $\sqrt{Ab}\ g^{(1)}(t=0)$ was made equivalent with the integrated excess scattered intensity

$$\sqrt{Ab}\ g^{(1)}(t=0) = \sqrt{Ab} \int_0^\infty G(\Gamma) d\Gamma = \sqrt{Ab} = <I> \tag{6}$$

If the CONTIN algorithm was running in equal spacing on the logarithmic scale and the linewidth distribution was normalized by the area, then the intensity of scattered light for each fraction Γ_i can be expressed as $G(\ln\Gamma_i)$ which is related to $G(\Gamma_i)$ by the relation

$$\int_{\Gamma_{min}}^{\Gamma_{max}} G(\Gamma) d\Gamma \qquad \text{(in equal spacing on } \Gamma \text{ scale)}$$

$$= \int_{\ln\Gamma_{min}}^{\ln\Gamma_{max}} G(\Gamma)\Gamma\ d\ln\Gamma \qquad \text{(in equal spacing on } \ln\Gamma \text{ scale)}$$

$$\text{or} \quad G(\ln\Gamma_i) = \Gamma_i G(\Gamma_i) \tag{7}$$

The next step is to transform $G(\Gamma_i)[\text{or } G(\ln\Gamma_i)]$ to the molecular weight distribution (MWD).

(c) Molecular weight distribution.

According to eq. (1), we have

$$R_{vv} \sim \frac{M_w P(\theta)}{(1 + 2A_2 M_w C)} \qquad (8)$$

In eq. (8), the term $M_w/(1 + 2A_2 M_w C)$ can be considered to be a bulk property, i.e., is independent of each fraction. In other words, the second virial coefficient is assumed to be a constant, independent of molecular weight in the polydiserse system under consideration. However, the form factor, $P(\theta)$ can play a very important role in the transformation of $G(\Gamma)$ into MWD.

$$\langle I(K) \rangle = \int f(M)M \, P(K,M)dM \qquad \text{(equal spacing on M scale)}$$

$$= \int f(M)M^2 \, P(K,M) \, d\ln M \qquad \text{(equal spacing on lnM scale)}$$

or

$$\langle I(K) \rangle \simeq \Delta \ln M \sum_i f(M_i)M_i^2 \, P(K,M_i) \qquad (9)$$

Here $f(M_i)$ is the weight fraction of molecular weight M_i using equal spacing in M scale and $P(K,M_i)$ is the form factor for the molecular weight M_i at the scattering vector K, and the final expression in eq. (9) approximates the MWD as a discrete distribution. If Γ_i is related to M_i by an empirical power law ($\Gamma \sim M^{-\alpha_D}$), $\Delta\ln\Gamma$ can be linked with $\Delta\ln M$ through a proportionality constant. In our case, the weight fraction of each M_i in lnM-space, $F_w(\ln M_i)$ has the form

$$F_w(\ln M_i) = f(M_i)M_i \sim \frac{\Gamma_i G(\Gamma_i)}{M_i P(K,M_i)} \qquad (10)$$

as $C \to 0$. In computing the molecular weight distribution, we also need the molecular weight dependence of $D(=\lim_{K\to 0} \Gamma/K^2)$ and of R_g. From laser light scattering characterization of unfractionated polymer samples of different molecular weight, we obtained

$$\overline{D}_z^0 = k_D M_w^{-\alpha_D} \qquad (11)$$

$$R_g = k_R M_w^{\alpha_R} \qquad (12)$$

By combining eq. (5) with eq. (11), we can imply a relation between the molecular weight, M , and the characteristic linewidth, Γ, at a finite scattering angle and a finite concentration

$$M_i = [\frac{\Gamma_i}{K^2 k_D (1 + fR_g^2 K^2)(1 + k_d C)}]^{1/\alpha_D}$$ (13)

$$= (const. \; \Gamma_i)^{1/\alpha_D}$$

In eq. (13), we have included the $(1 + fR_g^2 K^2)$ term without explicitly inserting eq. (12). It should be noted that at small scattering angles, $fR_g^2 K^2 \ll 1$. However, for broad MWD, the high molecular weight fraction often requires the $(1 + fR_g^2 K^2)$ correction term in eq. (13). In eq. (10), we may also express $P(K, M_i)$ as

$$P^{-1}(K, M_i) = 1 + \frac{k_R^2 M_i^{2\alpha_R} K^2}{3}$$ (14)

Thus, for the conversion of $F_w(lnM)$ from $G(\Gamma_i)$ we can use eqs. (10) and (14), while for the x-axis we have eq. (13) using lnM spacing.

4.2 Analysis of size exclusion chromatogram

The most widely accepted parameter defining the elution volume v in SEC is the hydrodynamic volume, (13) which for flexible coils of finite dilution is proportional to the product of the molecular weight M and the intrinsic viscosity $[\eta]$ ($\equiv \lim_{C \to 0}$ $(\eta - \eta_o)/(\eta_o C)$ with η_o being the solvent viscosity). Under fixed experimental conditions, the relation between $M[\eta]$ and elution volume has been shown to be independent of the nature of the structure of the polymer chain over a wide range with few exceptions. (14) Hence, the calibration curve $U(v) (= M[\eta])$ can be set to be proportional to R_η, with R_η being the viscometric hydodynamic radius of the polymer coil.

$$U(v) = M[\eta] \equiv \frac{10\pi N_A R_\eta^3}{3}$$ (15)

The calibration curve $U(v)$ can be determined from well-characterized fractions of linear polystyrene. Under the same fixed experimental conditions and for two different polymers

$$M_1[\eta]_1 = M_2[\eta]_2$$ (16)

at the same elution volume, i.e., the SEC detector cannot distinguish polymers having the same $[\eta]$ but different structures. If these two different polymers follow the Mark-Houwink equation over our range of interest, a simple substitution of $[\eta]$ by $[\eta] = K_\eta M^\alpha$ immediately leads to an equation which transforms the primary calibration curve (obtained using polymer 1) for use in other polymers as denoted by subscript 2.

$$\ln M_2 = \frac{1}{(1 + \alpha_2)} \ln \frac{k_{\eta_1}}{k_{\eta_2}} + \frac{(1 + \alpha_1)}{(1 + \alpha_2)} \ln M_1 \tag{17}$$

at the same elution volume. Therefore, a calibration curve $v = H(M)$, relating the logarithm of molecular weight of the eluant to the elution volume v for any polymer, can be obtained from the primary calibration curve and eq. (17). (15,16)

Next, in order to transform the chromatogram denoting a plot of concentration C versus elution volume v into a plot of the weight fraction, $f(M)$ versus the molecular weight M, both curves were normalized.

$$\int C(v)dv = \int f(M)dM = 1 \tag{18}$$

We set

$$f(M) = -C(v) \frac{dv}{dM} \tag{19}$$

with the negative sign accounting for the fact that high molecular weight fractions appear at low elution volumes. In $\ln M$-space we can express the weight fraction $F_w(\ln M)$ as

$$F_w(\ln M) = -C(v)M \frac{dv}{dM} \tag{20}$$

As illustrated by eqs. (19) and (20), we need a calibration curve relating the elution volume, v, with the molecular weight (M) in order to obtain a correct transformation from the chromatogram to the molecular weight distribution (MWD). In other words, the calibration curve in SEC provides the empirical information needed in the transform of the elution volume to the molecular weight (for the x-axis of MWD). If an empirical equation such as $v = b_1 + b_2 \ln M + b_3 (\ln M)^2$ is used for the calibration curve, the molecular weight (in $\ln M$-space) can be calculated from the chromatogram $C(v)$ according to the following procedure.

$$M_w = \int f(M)M dM = - \int C(v) \left(\frac{dv}{dM}\right) \cdot M^2 \, d\ln M$$

$$= - \int C(v) \left(\frac{b_2}{M} + \frac{2b_3 \ln M}{M}\right) M^2 \, d\ln M,$$

or

$$M_w \simeq - \Delta \ln M \sum_i C(v_i) M_i \, (b_2 + 2b_3 \ln M_i) \qquad (21)$$

which can be simplified to

$$M_w \simeq -b_2 \Delta \ln M \sum_i C(v_i) M_i \qquad (22)$$

if $b_3 = 0$. Thus, in a simple linear calibration curve (i.e. $b_2 < 0$, $b_3 = 0$), the molecular weight distribution in $\ln M$-space can be computed easily according to eq. (22) without any manipulation on the height of the chromatogram if the detector in SEC measures the weight concentration. It should also be noted that the universal molecular weight calibration curve for SEC works better with $[\eta]$ M_n. (17) We shall use M_w instead of M_n in our plots because the natural variable is $^w M$ from light scattering intensity measurements. Furthermore, most scaling relations of R_g and R_h versus M have M expressed in M_w.

Results and Discussion

1. Degree of Branching

    In order to estimate the degree of branching, we need information to establish a baseline concerning the behavior of linear PVAc in terms of the scaling relations $R_g = k_R M_w^{\alpha_R}$, $[\eta] = K_\eta M_w^\alpha$, etc., which are essentially available in the literature. (18,19) Only one sample (L_5) among our linear fractions was measured for comparison with the literature.

    The z-average radius of gyration, R_g, and the weight-average molecular weight, M_w, were determined by means of a Zimm plot as shown typically in Fig. 1. As values of the radius of gyration R_g and of the intrinsic viscosity $[\eta]$ for sampe L_5 are in better agreement with experimental data reported by Shultz, (18) we chose the Shultz data as our reference for linear PVAc. Figures 2 and 3 show log-log plots of radius of gyration and intrinsic viscosity $[\eta]$ as a function of molecular weight for both linear and branched polymers, respectively. As expected, $R_{g,b}(M) < R_{g,l}(M)$ and $[\eta]_b <$ $[\eta]_l$ with the subscripts b and l denoting branched and linear polymers. Furthermore, the effect increases with increasing degree of branching.

    Empirical equations for linear PVAc in MEK at $25^\circ C$ were established.

$$R_g = 1.28 \times 10^{-2} \, M_w^{0.59} \quad (nm) \qquad (23)$$

$$[\eta] = 1.29 \times 10^{-2} \, M_w^{0.71} \quad (1/kg) \qquad (24)$$

with M_w expressed in units of g/mol. In eq. (24), we used only

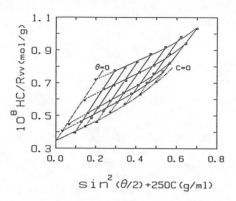

Figure 1. A Zimm plot for PVAc (L5) in MEK at 25°C.
$M_w = 1.58 \times 10^6$ g/mol. $A_2 = 2.82 \times 10^{-4}$ cm³ mol/g². $R_g = 56$ nm.
Note: mol = mole.

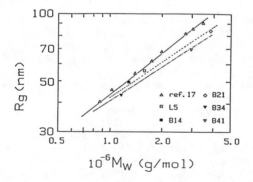

Figure 2. Log-log plot of R_g (nm) versus M_w (g/mol) for PVAc in MEK
at 25°C. The solid line denotes the linear PVAc, R_g (nm) =
1.28×10^{-2} $M_w^{0.59}$. The broken line (----) and the chain line (-·-·
-·-) denote B_2 and B_4 branching series, respectively. For the B_4
branching series, $R_{g,B4} \simeq 3.5 \times 10^{-2}$ $M_w^{0.51}$ (nm). $R_{g,b}(M) < R_{g,l}(M)$
for highly branched PVAc (e.g. B_{41}) where the subscripts b and l
denote branched and linear polymers, respectively.

the three data points in the higher molecular weight range of ref.
(18) and our L_5 data yielding $\alpha_\eta = 0.71$ which is an acceptable
exponent value.

The z-average translational diffusion coefficient at
infinite dilution, $\overline{D}{}_z^0$, could be determined by extrapolating $\overline{\Gamma}/K^2$
to zero scattering angle and zero concentration as shown typically
in Figs. 4 and 5. $\overline{D}{}_z^0$ is related to the effective hydrodynamic
radius, R_h by the Stokes-Einstein relation:

$$\overline{D}{}_z^0 = \begin{array}{c} \lim \\ K\to 0 \\ C\to 0 \end{array} (\frac{\overline{\Gamma}}{K^2}) = \frac{k_B T}{6\pi \eta_0 R_h} \tag{25}$$

with k_B being the Boltzmann constant. The viscometric
hydrodynamic radius, $R_\eta [\equiv (3M[\eta]/10\pi N_a)^{1/3}$ based on eq. (15)] of
linear PVAc (L_5) and the effective hydrodynamic radius, R_h [based
on eq. (25)], are in good agreement within experimental error
limits with $R_\eta \sim 43$ nm and $R_h \sim 42$ nm. We can estimate an
exponent α_h of the scaling relation $R_h = k_h M_w^{\alpha_h}$ by means of

$$\alpha_h = \frac{\alpha + 1}{3} \tag{26}$$

With $\alpha = 0.71$, we have $\alpha_h = 0.57$. By using L_5 as our calibration
standard, we can now determine the molecular weight dependence of
linear PVAc in MEK at 25°C. Figure 6 shows a log-log plot of R_h
and M_w for linear and branched PVAc in MEK at 25°C. The solid
line is denoted by

$$R_h = 1.22 \times 10^{-2} M_w^{0.57} \quad (nm) \tag{27}$$

$$\overline{D}{}_z^0 = 2.10 \times 10^{-8} M_w^{-0.57} \quad (m^2/sec) \tag{28}$$

with M_w expressed in g/mol. Again, at the same molecular weight,
$R_{h,b} < R_{h,1}$ and the effect increases with increasing degree of
branching. The degree of branching has been defined by several
types of ratios between the size of the linear chain and that of
the branched chain at the same molecular weight.

$$g = \frac{R_{g,b}^2}{R_{g,1}^2} \tag{29}$$

$$h = \frac{R_{h,b}}{R_{h,1}} \tag{30}$$

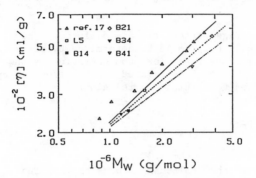

Figure 3. Log-log plot of $[\eta]$ (l/kg) versus M_w (g/mol) for PVAc in MEK at 25°C. The symbols are the same as those in Fig. 2. Solid line: $[\eta] = 1.29 \times 10^{-2} M_w^{0.71}$ (l/kg) for linear PVAc using only three data points in the high M_w range of ref. 17 and our L_5 data. Chain line: $[\eta]_{B4} \simeq 1.1 \times 10^{-1} M_w^{0.55}$ (l/kg) for the branched B_4 series. Note: l/kg = ml/g.

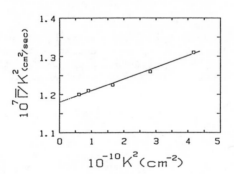

Figure 4. Plot of $\bar{\Gamma}/K^2$ versus K^2 for branched PVAc (B_{41}) in MEK at $C = 3.44 \times 10^{-4}$ kg/l and 25°C.

$$\lim_{K \to 0} \frac{\bar{\Gamma}}{K^2} = \bar{D}_z = 1.18 \times 10^{-11} \text{ m}^2/\text{sec}$$

$\bar{\Gamma}/K^2 = \bar{D}_z (1 + fR^2_{g,app} K^2)$ with $fR^2_{g,app} = 2.56 \times 10^{-16}$ m^2 and K expressed in m^{-1}.

$$G = \frac{[\eta]_b}{[\eta]_1} \qquad (31)$$

Table 1 lists the molecular parameters as well as g, h and G ratios of linear and branched PVAc in MEK at $25^\circ C$. Our most branched PVAc (B41) has $g \sim 0.70$, $G = 0.79$ and $h = 0.87$ suggesting that the branching effect is strongest for g but weakest for h. In fact, as g comes from the ratio of R^2 for linear and branched polymers, and h comes from the ratio of R_h^g (not R_h^2) for linear and branched polymers, the branching effect must necessarily be smaller for h than for g. Similarly, as G comes from the ratio of $[\eta] \sim R_\eta \sim R_h$ for linear and branched polymers, the branching effect must be bigger for G than for h. Zimm and Stockmayer (20) have derived equations relating g and the number of branch points (n_w) for randomly branched polymers having trifunctional or tetrafunctional branch points for both monodisperse and polydisperse systems. As our branched PVAc samples have trifunctional branch points, we can determine n_w values if the g values are known. (3) For $g \sim 0.70$, $n_w \sim 2.6$ suggesting 2-3 trifunctional branched points for our most branched PVAc (B41). Thus, for our branched PVAc samples, the overall degree of branching is relatively low.

An empirical expression between g and G was proposed by Park and Graessley (21) with

$$\ln g = 0.735 \ln G - 0.113 (\ln G)^2 \qquad (32)$$

According to eqs. (29)-(31), $G \sim h^3 \sim g^{3/2}$ provided that changes of R_g and R_h (or R_η) are comparable with respect to molecular weight. Figure 7 shows a plot of R_g/R_h versus M_w for linear and branched PVAc in MEK at $25^\circ C$. Within the degree of branching available for the PVAc fractions we were able to achieve the static (R_g) and dynamic (R_h) sizes for different molecular weight PVAc with different degrees of branching showing similar trends, i.e., the ratio R_g/R_h is not very sensitive to the degree of branching for PVAc polymers of the same molecular weight. Table 1 also lists values for $G^{1/3}$, h and $g_{1/2}$. It should be noted that in comparing the size ratios (h, $G^{1/3}$ and $g^{1/2}$) between branched and linear PVAc, the experimental error limits are fairly large ($\sim\pm10\%$) because we used estimates of sizes for linear PVAc of corresponding molecular weight by means of eqs. (23), (24) and (27). For branched polymers, the empirical scaling relation of the type such as eq. (23) should be modified at constant degree of branching since $R_{g,b} < R_{g,1}$ at high molecular weight and $R_{g,b} \sim R_{g,1}$ at low molecular weight. However, we shall ignore this subtle detail.

Sample B41 has the highest degree of branching as shown in Figs. 2, 3 and 6. Within the molecular weight range of our studies (between $\sim 5 \times 10^5$ g/mol and $\sim 1 \times 10^7$ g/mol) we may represent the molecular weight dependence of the parameters (R_g, R_h, $[\eta]$ and D_z) by assuming $B_{34} \approx B_{44}$. As B_{34} was in the molecular weight

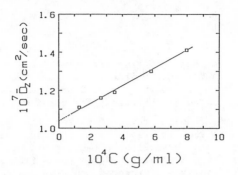

Figure 5. Plot of $\overline{D}_z$ versus concentration for branched PVAc (B_{41}) in MEK at 25°C. $\overline{D}_z^z = \overline{D}_z^0 (1 + k_d C)$ with $\overline{D}_z^0 = 1.04 \times 10^{-11}$ m^2/sec and $k_d = 420$ 1/kg.

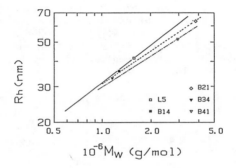

Figure 6.   Log-log plot of R_h(nm) versus M_w(g/mol) for linear and branched PVAc in MEK at 25°C.   The symbols are the same as those in Fig. 2.
R_h(nm) = $1.22 \times 10^{-2} M_w^{0.57}$ for linear PVAc of the solid line.
$R_{h,B4}^h$(nm) $\simeq 2.65 \times 10^{-2} M_w^{0.51}$ for B_4 series of the chain line.

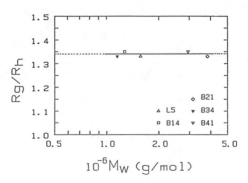

Figure 7. Plot of R_g/R_h versus M for linear and branched PVAc in MEK at 25°C.

Table 1. Molecular parameters of linear and branched PVAc in MEK at 25°C

Sample code	$M_w \times 10^{-6}$ (g/mol)	$A_2 \times 10^4$ (cm3 mol/g^2)	R_g (nm)	R_h (nm)	$[\eta]$ (1/kg)	R_η (nm)	$\dfrac{R_g}{R_h}$	$\dfrac{R_h}{R_\eta}$
L_5	1.58	2.82	56	42	312	43	1.33	0.98
B_{14}	1.28	2.84	50	36	252	37	1.35	0.97
B_{34}	1.16	2.96	44	33	243	36	1.33	0.92
B_{41}	2.96	2.29	70	52	402	57	1.35	0.91
B_{21}	3.83	2.42	84	63	563	70	1.33	0.90

Sample code	G (eq.31)	g (eq.29)	h (eq.30)	$g^{3/2}$	h^3	$G^{1/3}$	$g^{1/2}$
L_5	*	*	*	-	-	-	-
B_{14}	0.90	0.94	0.97	0.91	0.91	0.97	0.97
B_{34}	0.93	0.81	0.93	0.73	0.80	0.98	0.90
B_{41}	0.79	0.70	0.87	0.59	0.66	0.94	0.84
B_{21}	0.91	0.74	0.91	0.64	0.75	0.97	0.86

range where the size parameters begin to converge, the assumption should be acceptable for a qualitative estimate of the empirical scaling relations for branched PVAc. Then, we have

$$R_{g,B4} \simeq 3.5 \times 10^{-2} \, M_w^{0.51} \quad (nm) \tag{33}$$

$$R_{h,B4} \simeq 2.65 \times 10^{-2} \, M_w^{0.51} \quad (nm) \tag{34}$$

$$[\eta]_{B4} \simeq 1.1 \times 10^{-1} \, M_w^{0.55} \quad (1/kg) \tag{35}$$

$$\bar{D}_{z,B4}^{o} \simeq 2.1 \times 10^{-8} \, M_w^{-0.51} \quad (m^2/sec) \tag{36}$$

Eqs. (33), (34) and (35) are shown as dotted lines in Figs. 2, 3 and 6, respectively. Finally, we can estimate the degree of branching for the B_4 series which shows the most pronounced branching effect.

$$G \equiv \frac{[\eta]_b}{[\eta]_l} \simeq \frac{1.1 \times 10^{-1} \, M_w^{0.55}}{1.29 \times 10^{-2} M_w^{0.71}} \simeq 8.5 \, M_w^{-0.16} \tag{37}$$

2. Molecular Weight Distribution from Laser Light Scattering

We shall illustrate our analysis by using one normalized characteristic linewidth distribution $G(\Gamma_i)$ from the CONTIN method based on the experimental intensity-intensity time correlation function measured at scattering angle $\theta = 33^{\circ}$, concentration $C = 3.44 \times 10^{-4}$ g/ml in MEK at 25°C. With $fR_{g}^2 \sim 2.56 \times 10^{-1\text{f}} \, m^2$ and $k_d \sim 420$ 1/kg being determined from Figs. 4 and 5, respectively, we have

$$\bar{\Gamma}/K^2 = \bar{D}_z^o (1 + 2.56 \times 10^{-16} \, K^2)(1 + 420C) \quad (m^2/sec) \tag{38}$$

with K expressed in m^{-1}. By substituting eqs. (36) and (38) into eq. (13), each Γ_i was converted into the corresponding M_i and the weight fraction $F^i(lnM)$ in lnM spacing was computed from $G(\Gamma_i)$ by eqs. (10), (14) and (33). The molecular weight distribution of B_{41} sample is shown in Fig. 8. The weight average molecular weight M_w (~2.96×10^6 g/mol) obtained from the Zimm plot was in good agreement with the weight average molecular weight, M_w^{cal} (~3.0×10^6 g/mol) calculated from this distribution. The results are summarized in Table 2.

3. Molecular Weight Distribution from Gel Permeation Chromatography

Intrinsic viscosity $[\eta] = 444$ 1/kg for the branched PVAc B_{41}

Table 2.   Comparison of LLS and SEC results

	LLS	SEC $\alpha = 0.55$	$\alpha = 0.58$
$10^{-6}xM_w$ g/mol	3.0	2.7	2.7
$10^{-6}xM_n$ g/mol	2.3	2.2	2.2
M_w/M_n	1.30	1.23	1.23

$^*M_w$ ~2.96x10^6 g/mol from the Zimm plot.

sample in THF at 25 °C. As both THF and MEK are good solvents for PVAc, we assumed that the exponent of the Mark–Houwink equation for the unfractionated B_4 branched PVAc in THF would be the same as that in MEK at 25 °C. A comparison of hydrodynamic radii of linear and of branched PVAc in different solvents, as listed in Tables 3 and 4, suggests support for this assumption. With $\alpha = 0.55$ and $[\eta] = 444$ l/kg, a Mark–Houwink equation for B_4 branched PVAc in THF at 25 °C may be expressed as

$$[\eta] = 1.23 \times 10^{-1} M_w^{0.55} \quad (1/kg) \tag{39}$$

The Mark-Houwink equation for the linear polystyrene/THF system at 25°C was determined accurately by W. Graessley et al. We used their equation as follows:

$$[\eta] = 1.25 \times 10^{-2} M_w^{0.713} \quad (1/kg) \tag{40}$$

Our primary calibration curve obtained with narrow MWD linear polystyrene samples could now be used to determine the molecular weight distribution of branched PVAc (B_4) by means of eqs. (39), (40) and (17). The calibration curve for B_4 branched PVAc was not linear but showed a slight upturn at the low elution volume range as shown in Fig. 9. The calibration curve in Fig. 9 could be represented by

$$v = 48.81 - 3.474 \ln M_{PVAC} + 8.215 \times 10^{-2} (\ln M_{PVAC})^2 \quad (ml) \tag{41}$$

with M expressed in g/mol. The elution volume v could be converted to the corresponding molecular weight by the calibration curve and the height $C(v)$ of the chromatogram could be transformed into the weight fraction, $F_w(\ln M)$ by eqs. (20) and (41).

The molecular weight distribution obtained from SEC analysis was also shown in Fig. 8. In order to check the effect of the estimated exponent $\alpha(\sim 0.55)$ on molecular weight distribution for B_4 branched PVAc, we used another $\alpha(\sim 0.58)$ value to compute a new calibration curve as shown in Fig. 9. The two calibration curves almost overlapped with each other. The results are listed in Table 2. In both cases, we obtained the same weight-average molecular weight and the polydispersity index (M_w/M_n). Thus, we could confirm that in using a two-point (B_{24} and B_{41}) estimate for α, we have not introduced an appreciable error in the determination of molecular weight distribution of branched PVAc.

The two molecular weight distributions (MWD) from SEC/viscometry and laser light scattering (LLS), as shown in Fig. 8, agreed reasonably well over the entire molecular weight range. The weight average molecular weight M_w ($\sim 2.7 \times 10^6$ g/mol) calculated from MWD of SEC was a little smaller than that ($\sim 3.0 \times 10^6$ g/mol) from light scattering. The number average molecular weight ($\sim 2.2 \times 10^6$ g/mol) was the same for both cases. If we consider the

Table 3.   Comparison of hydrodynamic radii of linear and branched
PVAc in two different solvents at 25°C.

Hydrodynamic radius (nm)

Sample	THF (A)	MEK (B)	Ratio (A/B)
PVAc L_5	37.3	42.0	0.89
PVAc B_{41}	47.0	52.0	0.90

Ave. 0.90

Table 4. Comparison of hydrodynamic radii of linear polystyrene
in two different solvents at 25°C

Sample	$M_w \times 10^6$	Hydrodynamic radius (nm)		
		THF[a] (A)	Benzene[b] (B)	Ratio (A/B)
PS,linear	1.8	47.1	46.6	1.01
	0.88	29.8	31.4	0.95
	0.411	20.2	20.7	0.98
	0.160	11.9	12.3	0.97
			Ave.	0.98

---

a) From ref. 21

b) Calculated from the molecular dependence equation of
translational diffusion coefficients, $\overline{D}_z^o = (2.18 \pm 0.32) \times 10^{-8}$
$M_w^{-0.55 \pm 0.02}$ (m^2/sec) in ref. 22.

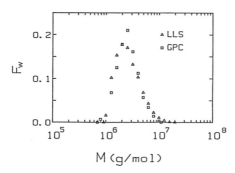

M (g/mol)

Figure 8. Molecular weight distribution of branched PVAc (B_{41}) as determined by laser light scattering (LLS) and size exclusion chromatography (SEC)/viscometry. A comparison of results by the two methods is listed in Table 2.

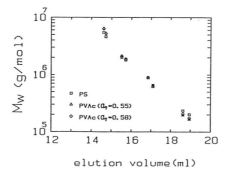

elution volume (ml)

Figure 9. SEC calibration curves for linear polystyrene (PS) and B_4 branched PVAc listed in Table 2.

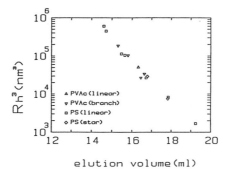

elution volume (ml)

Figure 10. A universal plot of $\log(R_h^3)$ versus elution volume (v) in THF at 25°C. Data are listed in Table 5.

Table 5.   Hydrodynamic radius (R_h) and elution volumes (v) of
several different polymers in THF at 25°C

Polymer	Structure	$M_w(\times 10^{-6}$ g/mol)	R_h (nm)	v (ml)
polystyrene	linear	0.160	11.9[a]	19.25[b]
		0.411	20.2[a]	17.85[b]
		0.88	29.8[a]	16.72[b]
		1.80	47.1[a]	15.64[b]
		2.00	48.4[c]	15.47
		4.6	76.5[c]	14.71
		5.5	84.4[c]	14.60
polystyrene	12-arm star	0.79	19.6[d]	17.83
		2.1	30.8[d]	16.78
polyvinyl acetate	linear L_5	1.58	37.0	16.32
	branch B_{41}	2.96	47.0	15.80
	B_{34}	1.16	29.7[e]	16.45
	B_{21}	3.83	56.7[e]	15.30
	B_{14}	1.28	32.4[e]	16.63

a) From ref. 21.

b) Interpolated from our universal calibration curve of standard
linear polystyrene samples.

c) Calculated from the equation $\overline{D}_z^0 = (2.18\pm0.32)\times10^{-8}\,M_w^{-0.55\pm0.02}$
m²/sec of ref. 22 and the ratio constant (0.98) shown in
Table 4.

d) Converted from our experimental R_h in benzene with a ratio
constant of 0.98

e) Converted from our experimental R_h in MEK with a ratio
constant of (0.90) shown in Table 3.

limits of resolution for our SEC columns in the high molecular weight range and the estimated calibration curve from linear polystyrene as well as the strong sensitivity of light scattering for high molecular weight fractions, a discrepancy (~10%) in M_w between SEC and LS is quite acceptable indeed. Thus, we note that laser light scattering could not only determine the degree of branching, but also the MWD of branched polymers without some of the complications by the SEC/viscometry method.

Conclusion

Intrinsic viscosity is proportional to the hydrodynamic volume while dynamic light scattering yields a translational diffusion coefficient which is inversely proportional to the hydrodynamic radius. Thus the size ratio $G(\equiv [\eta]_b/[\eta]_1)$ is more sensitive to the degree of branching than $h(\equiv R_{h,b}/R_{h,1})$. If we assume that an empirical scaling relation of the type $y = kM^\alpha$ is applicable for an unfractionated branched polymer, the exponent α and the proportionality constant k can be estimated based on experimental measurements of y for the branched polymer and the molecular weight dependence of y for the linear polymer under the same conditions. It is important to note that the SEC/viscometry technique requires a knowledge of the empirical relation $y = kM^\alpha$, in a similar way as required by LLS. However, as LLS is less sensitive to the degree of branching, the approximation in using an overall average α exponent in the diffusion coefficient/molecular weight transform should hold reasonably well for most practical purposes, at least for PVAc.

Finally, instead of $M[\eta]$ versus elution volume, we can construct a new universal calibration curve using $R_h^3 = (k_B T/6\pi\eta \bar{D}_z^o)^3$ for the y-axis as shown in Fig. 10. The potential for combining LLS with SEC is obvious. If we can perform LLS measurements on just a few fractions from SEC, and have knowledge on behavior of the linear polymer, we can determine M_w, MWD, and the degree of branching on those fractions, as well as other static and dynamic properties accessible by LLS.

Literature Cited

1. Small, P.A. Adv. Polym. Sci. 1975, 18, 1.
2. Park, W.S.; Graessley, W.W. J. Polym. Sci., Polym. Phys. Ed. 1977, 15, 85.
3. Drott, E.E.; Mendelson, R.A. J. Polym. Sci. (A-2), 1970, 8, 1361, 1373.
4. Wild, L.; Ranganath, R.; Ryle, T. J. Polym. Sci. (A-2), 1971, 9, 2137
5. Kraus, G.; Stacy, C.J. J. Polym. Sci. (A-2), 1972. 10, 657.
6. Tung, L.H. J. Polym. Sci. 1971, 9, 759.
7. Dietz, R.; Francis, M.A. Polymer, 1979, 20, 450.
8. Chu, B.; Onclin, M.; Ford, J.R. J. Phys. Chem. 1984, 88, 6566.
9. Brandrup, J.; Immergut, E.H., eds. "Polymer Handbook", John Wiley & Sons, New York, 1975.

10. Chu, B. "Laser Light Scattering", Chaps. 6,8, Academic Press, New York, 1974.
11. Stockmayer, W.H.; Schmidt, M.W. Pure Appl. Chem. 1982, 54, 407; Macromolecules 1984, 17, 509.
12. Provencher, S.W. Biophys. J. 1976, 16, 27; J. Chem. Phys. 1976, 64, 2772; Makromol. Chem., 1979, 180, 201.
13. Grubisic, A.; Rempp, P.; Benoit, H. J. Polym. Sci. B, 1967, 5, 753.
14. Pannell, J. Polymer 1972, 13, 277.
15. Coll H.; Gilding, D.K. J. Polym. Sci. A2 1970, 8, 89.
16. Atkinson, C.M.L.; Dietz, R. European Polym. J. 1979, 15, 21.
17. Hamielec, A.E.; Ouano, A.C. J. Liq. Chromatography 1978, 1, 111; Hamielec, A.E.; Ouano, A.C.; Nebenzahl, L.L. ibid. 1978, 11, 527.
18. Shultz, A.R. J. Am. Chem. Soc., 1954, 76, 3422.
19. Matsumoto, M.; Ohyanagi, T. J. Polym. Sci. 1960, 46,441.
20. Zimm, B.H.; Stockmayer, W. H. J. Chem. Phys. 1949, 17, 1301.
21. Park, W.S.; Graessley, W.W. J. Polym. Sci., Polym. Ed. 1977, 15, 71.
22. Mandema,W.; Zeldenrust, H. Polymer 1977, 18, 835.
23. Adam, W.; Delsanti, M. Macromolecules 1977, 10, 1229.

RECEIVED March 30, 1987

Chapter 15

Generalized Intrinsic Viscosity Relations for Copolymers and Higher Multispecies Polymers

for Size Exclusion Chromatographic Universal Calibration

Robert A. Mendelson

Monsanto Company, 730 Worcester Street, Springfield, MA 01151

An appropriate formalism for Mark-Houwink-Sakurada (M-H-S) equations for copolymers and higher multispecies polymers has been developed, with specific equations for copolymers and terpolymers created by addition across single double bonds in the respective monomers. These relate intrinsic viscosity to both polymer MW and composition. Experimentally determined intrinsic viscosities were obtained for poly(styrene-acrylonitrile) in three solvents, DMF, THF, and MEK, and for poly(styrene-maleic anhydride-methyl methacrylate) in MEK as a function of MW and composition, where SEC/LALLS was used for MW characterization. Results demonstrate both the validity of the generalized equations for these systems and the limitations of the specific (numerical) expressions in particular solvents.

Determination of the dilute solution intrinsic viscosities of homopolymers and the relating of these quantities to molecular weight was the first method of polymer molecular weight characterization and remains a principal method. The relationship used is generally called the Mark-Houwink-Sakurada (M-H-S) equation, and the constants of the equation are specific to each polymer-solvent system. These M-H-S equations have extensive usefulness throughout fundamental and applied polymer research. Definition of appropriate equations has become even more important with the advent of size exclusion chromatography (SEC) as a means of completely characterizing the molecular weight distributions (MWD) of polymers because

0097-6156/87/0352-0263$06.00/0
© 1987 American Chemical Society

of the recognition that SEC systems may be calibrated by
the so-called method of "universal calibration"(1). This
method is based on the observation(2) that the product of
the intrinsic viscosity times molecular weight (hydrodynamic volume) is a unique function of the elution volume
of an SEC column set for a wide variety of polymers in a
given solvent. Thus, the fact that well characterized
narrow MWD standards for calibration are available for
only a few polymeric species (e.g., polystyrene, polymethyl methacrylate) no longer limits the calibration of
SEC systems. These standards may be used to create a
universal calibration function applicable to essentially
all polymers which take on a random coil configuration in
solution. While application of this universal calibration for a particular polymer requires knowledge of
the M-H-S equation for that polymer in the chromatographic solvent, this has in general presented no major
problem for homopolymer characterization. However, the
case of copolymers and higher multispecies polymers is
more complex. Here it is expected that the intrinsic
viscosity is a function of both molecular weight and
polymer composition, and generalized M-H-S equations to
represent this complex functionality have not been used
extensively. Rather, the practice in the case of copolymers has been to evaluate the molecular weight dependence
of intrinsic viscosity at some fixed copolymer composition and to use this relationship, per se.

In this paper a generalized approach is presented to
the derivation of M-H-S equations for multispecies polymers created by addition polymerization across single
double bonds in the monomers. The special cases of
copolymers and terpolymers are derived. This development
is combined with experimental results to evaluate the
numerical parameters in the equations for poly(styrene-
acrylonitrile) (SAN) in three separate solvents and for
poly(styrene-maleic anhydride-methyl methacrylate)
(S/MA/MM) in a single solvent. The three solvents in the
case of SAN are dimethyl formamide (DMF), tetrahydrofuran
(THF), and methyl ethyl ketone (MEK); and the solvent for
S/MA/MM is MEK.

Theoretical Treatment

We attempt here to develop a mathematical expression for
the dependence of the dilute solution intrinsic viscosity
of multispecies polymers on both molecular weight and
polymer composition with some broad degree of generality
and to particularize the result for the specific cases of
copolymers and terpolymers such as SAN and S/MA/MM. The
details of the derivation are specific to polymers resulting from addition polymerization across a single double
bond in each monomer unit. In principle the approach may
be expanded to other schemes of polymerization so long as

the number of carbon atoms in the backbone chain can be related to a simple measure of composition. Indeed, Kruse and Padwa(3) have treated the specific case of styrene-butadiene copolymers, where there are two poly-merizable double bonds in the butadiene monomer (they assume all 1,4-addition). A second restriction on the current work is that of applicability to linear polymers, only. The problem to be addressed is that of developing a continuous relation between intrinsic viscosity and molecular weight and composition, while retaining sufficient simplicity to allow experimental definition of the equation's parameters for specific polymer-solvent systems. Three factors, which are consequences of polymer compositional change must be addressed. First, since the different monomer species' molecular weights are generally different, the relationship between polymer molecular weight and chain size is a function of composi-tion. Second, introduction of different monomer species into the backbone may alter the unperturbed coil dimen-sions; and, third, the solvent-polymer interaction may vary with polymer composition. It should be noted that these issues have been addressed in a very different manner by Goldwasser and Rudin(17).

The starting point is the hypothesis that the intrinsic viscosity is a direct function of the number of carbon atoms in the backbone chain. This addresses the first of the above considerations and is consistent with the homopolymer case and with the approach of Kruse and Padwa(3). The specific form of the functional dependence of intrinsic viscosity on the number of backbone carbon atoms is given by

$$[\eta] = JZ^{\beta} \tag{1}$$

where Z is the number of carbon atoms in the backbone, and J and β include the averaged conformational and excluded volume terms. Thus, J and β are analogous to the classical K and α terms in the homopolymer M-H-S equation. In this derivation it is assumed that J and β are constant over some importantly wide range of compo-sitional change. Thus, the second and third concerns stated earlier are initially addressed by assuming that the balance of segment-segment and solvent-segment inter-actions is essentially invariant over some wide composi-tion range. As will become apparent, this assumption is readily testable experimentally for specific systems.

For the case of polymerization by addition across single double bonds in each monomer, Z may be written as

$$Z = 2M_p \ \Sigma w_i/M_i \qquad (i = 1 -- n) \tag{2}$$

where M_p is the polymer molecular weight and w_i and M_i are the weight fraction and molecular weight of the ith monomer type in the polymer of n monomer types. Thus, the general equation for the intrinsic viscosity of an n-species polymer follows immediately as

$$[\eta] = 2^\beta J \, M_p^{\ \beta} \, (\Sigma w_i/M_i)^\beta \qquad (i = 1 -- n) \qquad (3)$$

where the M_i's are known, M_p and w_i are analytically determinable, and J and β must be experimentally determined for any specific polymer in a specific solvent. For the case of copolymers, i = 1,2 and

$$[\eta] = 2^\beta J \, M_p^{\ \beta} \, (w_1/M_1 + [1-w_1]/M_2)^\beta \qquad (4)$$

Equation (4) may be rearranged in a number of ways. A particular form lending simplicity to the final expression is

$$[\eta] = \{2^\beta J[(M_2-M_1)/M_1 M_2]^\beta\} \, M_p^{\ \beta} \, \{w_1 + M_1/(M_2-M_1)\}^\beta \qquad (5)$$

where the term in the first brackets is expected to be a constant involving J and β and the two monomer molecular weights, and the composition term, $\{w_1 + M_1/(M_2-M_1)\}$, includes a constant which is uniquely determined by the monomer molecular weights. In the specific case of SAN, taking AN as monomer-1, M_1 = 53.1 and M_2 = 104.1, eq.(5) yields

$$[\eta] = (1.845 \times 10^{-2})^\beta J \, M_p^{\ \beta} \, (w_{AN} + 1.041)^\beta \qquad (6)$$

where J and β must be determined experimentally in particular solvents. The terpolymer case follows readily from eq.(3):

$$[\eta] = 2^\beta J \, M_p^{\ \beta} \, (w_1/M_1 + w_2/M_2 + w_3/M_3)^\beta \qquad (7)$$

Again, the terms may be rearranged in a number of ways; the particular form chosen here is

$$[\eta] = (2/M_3)^\beta J M_p^{\ \beta} \{w_1(M_3-M_1)/M_1 + w_2(M_3-M_2)/M_2 + 1\}^\beta \qquad (8)$$

In the specific case of S/MA/MM, letting MA be monomer-1, MM be monomer-2, and styrene be monomer-3, M_1 = 98, M_2 = 100, and M_3 = 104. Introducing these values into eq.(8) yields

$$[\eta] = (1.923 \times 10^{-2})^\beta J M_p^{\ \beta} (6.122 \times 10^{-2} w_{MA} + 4.000 \times 10^{-2} w_{MM} + 1)^\beta \qquad (9)$$

Since M_2 and M_3 are very similar, the S/MA/MM terpolymer might be treated as a copolymer (S/MA), lumping methyl methacrylate and styrene together, and rewriting eq.(5) as

$$[\eta] = (1.177 \times 10^{-3})^\beta \, J \, M_p^{\ \beta} \, (w_{MA} + 16.33)^\beta \qquad (10)$$

Again, J and β must be determined experimentally for specific solvents. These predictions will be examined in the Results section of this paper.

Experimental

The polymer samples studied here fall into three distinct categories. Data from two sample populations have been combined in the SAN copolymer study. A group of SAN materials having compositions ranging from 42 (wt)% AN to 82% AN were polymerized and characterized quite some time ago (1972), with intrinsic viscosities determined only in DMF. Very recently, a second group of SAN's with compositions from 5 (wt)% to 48% AN, as well as one sample of polystyrene (0% AN), were polymerized and characterized, with intrinsic viscosities determined in DMF, THF, and MEK. These two populations are differentiated in the Results section by the designations "old data" and "new data". The third category of samples is that of S/MA copolymers and S/MA/MM terpolymers, with intrinsic viscosities measured only in MEK.

In all cases, intrinsic viscosities were measured at 25°C in constant temperature baths controlled to ±0.1°C or better, using suspended level Ubbelohde dilution viscometers with solvent flow times of at least 100 sec.. No kinetic energy corrections were made. Solution flow times were measured at four concentrations for each sample, and intrinsic viscosities were obtained from the classical double extrapolation of η_{sp}/c vs. c and $(\ln \eta_r)/c$ vs. c to a single intercept value. Concentration ranges were varied somewhat with the molecular weights of the samples, but were chosen such that both functions were straight lines in all cases.

The characterized molecular weights used in this investigation were in all cases weight average values, M_w, obtained by SEC. However, the SEC method varied, as might be expected, both with the nature of the polymer characterized and with the time at which the measurements were made. Thus, the "new" polystyrene and SAN samples were characterized using a dual detection SEC/LALLS (Waters 150C SEC/LDC-Milton Roy KMX-6 low angle laser light scattering detector) system($\underline{4},\underline{5}$). This system uses a differential refractive index (DRI) detector as the concentration detector and LALLS to measure excess scattering intensity (proportional to molecular weight). As is well known by now, the use of the light scattering detector permits the direct measurement of molecular weight as a function of retention volume, without recourse to the necessity of any form of column calibration. For all characterizations in this system the column set consisted of a 10^3, 10^4, 10^5, 10^6 A assembly (either Waters μ-Styragel or ASI Ultragel crosslinked polystyrene). The chromatographic solvent was THF containing 250ppm antioxidant (BHT). Refractive index

increment (dn/dc) values were independently determined
using a precision differential refractometer (LDC/Milton
Roy KMX-16). Although the data are not discussed here,
dn/dc was found to be linear with AN content, as might be
expected. From these data it was relatively straightforward to assess the expected error of using a sample
average dn/dc to calculate the individual MW's of the
eluants from light scattering, assuming the worst case of
combined compositional heterogeneity and compositional
correlation with molecular size in a given sample. Thus,
a variation of +5% AN content (absolute) about the mean
of the sample, if correlated with elution volume, would
be expected to result in approximately +5% error in the
sample Mw; no correlation of compositional heterogeneity
with elution volume, of course, eliminates the error.
Similarly, errors in the DRI measurement of concentration
due to compositional heterogeneity correlated with elution volume were viewed as potentially relatively small.
Of course, additional detectors to follow possible compositional drift(18) would eliminate these potential
sources of error, but they were not used in this work.
The light sources for both LALLS and the KMX-16 were
He-Ne lasers with 6328 A incident wavelength. The
accuracy of the SEC/LALLS was checked by determining the
molecular weights of a number of narrow MWD standards,
with excellent results.

By contrast, the "old" SAN's having compositions
from 42% to 82% AN were characterized(6) in 1972 using a
single detector SEC system with a column set consisting
of four porous glass bead columns (Porasil) calibrated
against osmometrically determined M_n values for several
of the samples studied. That is, the "Q-factor"
calibration obtained from polystyrene standards was
modified to force correct SEC evaluation of M_n for a
number of high AN SAN samples of known M_n. In the case
of this group of samples, the chromatographic solvent was
DMF containing 0.05M LiBr as an electrolyte.

In the case of the third category of samples, S/MA
and S/MA/MM, the M_w values were determined by A. S.
Kenyon(7) using a SEC/LALLS system similar to that
described earlier, with μ-Styragel columns (again, 10^3,
10^4, 10^5, 10^6 A). The chromatographic solvent in this
case was DMF containing maleic acid as an electrolyte,
and values of dn/dc were measured as described above for
the "new" SAN work. In some cases the SEC/LALLS-derived
M_w's were checked against those obtained by "static"
light scattering measurement at a series of
concentrations, with good agreement.

Average copolymer compositions of SAN samples were
determined by elemental analysis, yielding weight percent
acrylonitrile in the polymer. Compositions of S/MA and
S/MA/MM were determined by sequential hydrolysis and
pyridine titration to obtain maleic anhydride content and
by infrared analysis for methyl methacrylate content.

Results and Discussion

The combined characterization results for all of the SAN samples (including the one polystyrene sample), i.e., weight fraction of AN mer units (w_{AN}), M_w, and [η] in each solvent, are summarized in Table I. Where blanks exist in the [η] columns, the intrinsic viscosity was not measured for that particular sample in that particular solvent. Equation (6) predicts that a plot or regression of log[η] vs. log[$M_w(w_{AN} + 1.041)$] should test the validity of the relationship and, if valid, should yield β and J from the slope and intercept, respectively. Therefore, the final column of Table I gives the calculated values of [$M_w(w_{AN} + 1.041)$]. To test the validity of eq.(6), the large body of SAN [η] data for DMF as the solvent was initially plotted as indicated above with certain omissions. Because DMF is known to be a poor solvent for polystyrene relative to its solvent power for SAN's of reasonably high level of AN, the polystyrene sample and those SAN samples containing 10% AN or less were omitted. This plot, including both "old" and "new" data, is shown here as Figure 1, demonstrating a good straight line fit to the data (over the compositional range, 15-82% AN, and the M_w range, 45,000-800,000), and confirming the validity of eq.(6) for this copolymer-solvent system. It also indicates no systematic behavior difference between the old and new data sets. Linear least squares fitting of the data yields

$$[\eta]_{25°C,DMF} = 1.86 \times 10^{-4} (w_{AN} + 1.041)^{0.690} M_w^{0.690} \qquad (11)$$

where the units of [η] and M_w here and throughout this paper are dl/g and daltons, respectively. The R^2 correlation coefficient for the logarithmic form of eq.(11) is 0.967, which must be considered quite good considering the different time frames of the data and the accumulated inherent errors of all of the measurements involved. Thus, from eqs.(6) and (11), $\beta = 0.690$ and J = 2.93×10^{-3}.

Returning to the omitted AN $\leq$ 10% data, Figure 2 repeats the data in Figure 1, plus the additional five points. It is apparent that all five points systematically fall below the line representing eq.(11). This is readily explained in terms of a rapid decrease in the thermodynamic solvent power of DMF with decreasing AN content of the copolymer at some composition below 15% AN. This very interesting result addresses the third concern expressed early in the Theoretical section and suggests that polymer-solvent interaction effects on coil dimensions as a function of copolymer composition are relatively abrupt, rather than continuous and gradual. This is necessary for the modified M-H-S equation expressed in eq.(11) to be valid over any significantly wide polymer compositional range, and it is in contrast

TABLE I. Molecular Structure - [η] Data for SAN Samples

Sample	w_{AN}	$M_w \times 10^{-3}$	[η] (dl/g)			$M_w(w_{AN}+1.041)$
			DMF	THF	MEK	
N - PS 10	0	280	0.55	0.909	0.504	2.91 X 10^5
N - SAN 200 L	0.058	405	1.068	1.489	1.035	4.45 X 10^5
N - SAN 201 L	0.069	84.8	0.327	0.469	0.334	9.41 X 10^4
N - SAN 202 L	0.100	411	1.175	1.561	1.053	4.69 X 10^5
N - SAN 203 L	0.100	84.5	0.436	0.549	0.409	9.64 X 10^4
N - SAN 204 L	0.147	498	1.525	1.783	1.268	5.92 X 10^5
N - SAN 205 L	0.149	141	0.637	0.733	0.535	1.68 X 10^5
N - SAN 206 B	0.185	263	1.033	1.116		3.22 X 10^5
N - SAN 207 P	0.215	177	0.735	0.785	0.621	2.22 X 10^5
N - SAN 208 P	0.232	143	0.678			1.82 X 10^5
N - SAN 209 P	0.261	125	0.641	0.659	0.569	1.63 X 10^5
N - SAN 210 P	0.268	251	1.063	1.087	0.858	3.29 X 10^5
N - SAN 211 P	0.326	86.8	0.591	0.594	0.478	1.19 X 10^5
N - SAN 212 P	0.330	113	0.693	0.691	0.549	1.55 X 10^5
N - SAN 213 L	0.415	96.5	0.713	0.470	0.508	1.41 X 10^5
O - SAN 214 L	0.423	128	0.785			1.87 X 10^5
O - SAN 215 L	0.450	78.4	0.615			1.17 X 10^5
N - SAN 216 L	0.481	289	1.801	0.850	1.109	4.40 X 10^5
N - SAN 217 L	0.483	804	3.274	1.287	1.889	1.23 X 10^6
O - SAN 218 L	0.619	62.7	0.543			1.04 X 10^5
O - SAN 219 L	0.620	44.8	0.447			7.44 X 10^4
O - SAN 220 L	0.621	101	0.714			1.68 X 10^5
O - SAN 221 L	0.621	98.3	0.760			1.63 X 10^5
O - SAN 222 P	0.632	100	0.797			1.67 X 10^5
O - SAN 223 L	0.635	45.5	0.457			7.63 X 10^4
O - SAN 224 L	0.646	73.2	0.568			1.23 X 10^5
O - SAN 225 L	0.654	115	0.835			1.95 X 10^5
O - SAN 226 L	0.666	151	1.067			2.58 X 10^5
O - SAN 227 P	0.666	93.0	0.762			1.59 X 10^5
O - SAN 228 L	0.672	90.9	0.748			1.56 X 10^5
O - SAN 229 P	0.679	90.5	0.803			1.56 X 10^5
O - SAN 230 P	0.68	113	0.880			1.95 X 10^5
O - SAN 231 L	0.697	115	0.935			2.00 X 10^5
O - SAN 232 L	0.718	76.6	0.667			1.35 X 10^5
O - SAN 233 L	0.723	149	1.102			2.63 X 10^5
O - SAN 234 L	0.739	149	1.125			2.65 X 10^5
O - SAN 235 P	0.823	90.5	0.668			1.69 X 10^5

N = new data
O = old data

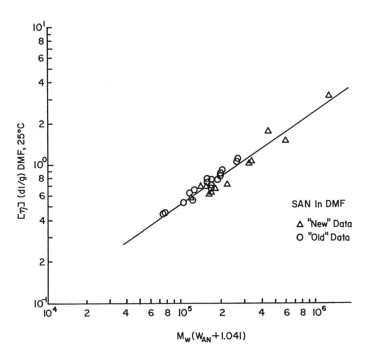

Figure 1.   Intrinsic viscosities of SAN samples in DMF solvent vs. combined structure parameter, $M_w(w_{AN} + 1.041)$, for samples with w_{AN} = 0.15 - 0.82.

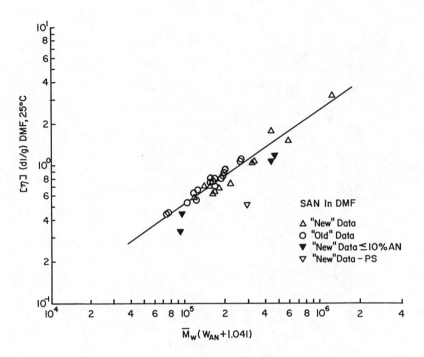

Figure 2.   Intrinsic viscosities of SAN samples in DMF
solvent vs. combined structure parameter,
$M_w(w_{AN} + 1.041)$, extending composition range to
include samples with $w_{AN} = 0 - 0.82$.

to the results of Goldwasser and Rudin([17]) for
poly(styrene-methyl methacrylate) copolymers where both
M-H-S constants were reported to be continuously varying
functions of composition.

The less extensive intrinsic viscosity data for SAN
in THF and in MEK (all "new" data) are given in Table I
and are plotted in Figure 3, again following from eq.(6),
this time including the polystyrene and low AN samples.
It is immediately evident for the THF solvent case that
the polystyrene and low AN SAN samples fit well, but that
the three samples at 42% and 48% AN fall well below the
best line through the remainder of the data. Thus, as
might be expected from a knowledge that very high AN SAN
is not soluble in THF, a rapid decrease in solvent power
may be deduced to occur in the copolymer composition
region somewhere between 33% and 42% AN. Omitting the
three high AN data points, a linear least squares best
fit of $\log[\eta]$ vs. $\log\{M_w(w_{AN} + 1.041)\}$ was calculated for
the $[\eta]_{THF}$ data and, converted from the logarithmic form,
is given by

$$[\eta]_{25°C,THF} = 1.98 \times 10^{-4} (w_{AN} + 1.041)^{0.681} M_w^{0.681} \quad (12)$$

with an R^2 correlation coefficient of 0.986 for the
logarithmic form covering the applicable composition
range from 0% to ca. 35% AN. Thus, from eqs.(6) and
(12), $\beta = 0.681$ and $J = 3.00 \times 10^{-3}$.

Turning to the SAN in MEK data, also plotted in
Figure 3, it is apparent that the higher AN SAN sample
data (42% and 48% AN) fit on the general straight line
representation, as do the two samples at 10% AN and one
of the two samples at 6-7% AN. However, the other low AN
(7%) and polystyrene fall below the line, suggesting
reduced solvent power in this composition region. The
linear least squares fit of the logarithmic form was
again calculated, in this case omitting the polystyrene
and the four lowest AN level SAN samples from the
calculation. Converting to the exponential form, this
yields

$$[\eta]_{25°C,MEK} = 3.30 \times 10^{-4} (w_{AN} + 1.041)^{0.619} M_w^{0.619} \quad (13)$$

with an R^2 correlation coefficient of 0.995 for the
logarithmic form of the equation. Here the applicable
composition range is somewhat ambiguous; it clearly
extends to at least 48% AN on the high side and to less
than 10% AN on the low side (perhaps to even lower AN
content). A conservative definition of the applicable
copolymer composition range of eq.(13) is 10-48% AN.
From eqs.(6) and (13), $\beta = 0.619$ and $J = 3.90 \times 10^{-3}$.

There are a variety of M-H-S equations in the
literature for polystyrene and several at fixed compositions for SAN copolymers, and there is considerable
variation in the values of K and α given. However, it is

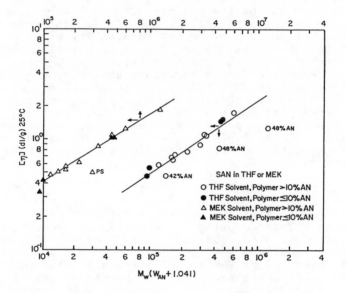

Figure 3.   Intrinsic viscosities of SAN samples in THF
solvent and in MEK solvent vs. combined structural
parameter, $M_w(w_{AN} + 1.041)$, for samples with
$w_{AN} = 0 - 0.48$.

possible to make some comparisons with the results of the current work. Table II summarizes values of K and α ($\alpha \equiv \beta$) for polystyrene in MEK and SAN in MEK and in DMF at fixed copolymer compositions evaluated from the current work and from several references. Based on the already noted possibility that eq.(13) may be extended below the stated range of applicability, the comparison for polystyrene in MEK appears reasonably justified. However, no such comparison for polystyrene in DMF appears justified based on the clear change in solvent power at low AN contents. The author is unaware of any appropriate literature M-H-S equations for SAN in THF; however, a number of references are available for polystyrene in THF, and several are given in Table II. Good agreement is observed with Shimura(8) for polystyrene in MEK, as is relatively good agreement with other investigators whose results fall on either side of the current results. For SAN in MEK current results are in good agreement with Shimura at the two compositions available from his work, but are at variance with the results of Lange and Baumann(12,13) whose value of α changes drastically over the composition range, where current data give a constant α (or β). By contrast, in DMF the Lange and Baumann results are in good agreement with the current work and show a relatively constant α with changing composition, while the result of Shimura at the one available composition gives a considerably higher value of α. In the case of THF the only comparison is for polystyrene, where the exponent from the current work appears somewhat low compared to the specific references cited.

Finally, we examine the terpolymer case using data obtained for S/MA/MM terpolymers, as well as S/MA copolymers. Table III summarizes the measured weight fractions of maleic anhydride (w_{MA}) and of methyl methacrylate (w_{MM}) mer units in the polymers, the M_w's, and the intrinsic viscosities in MEK at 25°C. Also, the terms, $\{M_w(6.12 \times 10^{-2}w_{MA} + 4.00 \times 10^{-2}w_{MM} + 1)\}$ and $\{M_w(w_{MA} + 16.33)\}$, as predicted from eqs.(9) and (10), are tabulated. In Figure 4 all of these data are plotted both according to eq.(9) and according to eq.(10). It may be seen that both forms result in good straight line fits to the data. Again, linear least squares fitting of the logarithmic form of the data was performed. For the terpolymer expression eq.(9), the resulting equation in exponential form is

$$[\eta]_{25°C,MEK} =$$
$$3.10 \times 10^{-4}(6.12 \times 10^{-2}w_{MA} + 4.00 \times 10^{-2}w_{MM} + 1)^{0.608}M_w^{0.608} \quad (14)$$

with an R^2 correlation coefficient (logarithmic form) of 0.973, and yielding $\beta = 0.608$ and $J = 3.42 \times 10^{-3}$. Treating the data in copolymer form (assuming methyl methacrylate may be neglected and treated as additional styrene mer units, i.e., eq.(10)) yields

TABLE II.   Comparison of M – H – S Parameters at fixed AN Content with
Literature Results

Solvent	w_{AN}	TEMP. (°C)	K (dl/g)	α	Reference
MEK	0	30	2.3×10^{-4}	0.62	Shimura (8)
MEK	0	25	3.9×10^{-4}	0.58	Outer, Carr, Zimm (9)
MEK	0	25	1.95×10^{-4}	0.635	Oth, Desreux (10)
MEK	0	25	3.05×10^{-4}	0.60	Bawn, et al (11)
MEK	0	25	3.38×10^{-4}	0.619	This work
MEK	0.10	30	1.5×10^{-4}	0.70	Lange, Baumann (12)
MEK	0.10	25	3.58×10^{-4}	0.619	This work
MEK	0.24	30	3.6×10^{-4}	0.62	Shimura (8)
MEK	0.24	30	2.5×10^{-4}	0.67	Baumann, Lange (13)
MEK	0.24	25	3.85×10^{-4}	0.619	This work
MEK	0.46	30	5.3×10^{-4}	0.61	Shimura (8)
MEK	0.50	30	9.8×10^{-4}	0.56	Lange, Baumann (12)
MEK	0.50	25	4.31×10^{-4}	0.619	This work
DMF	0.24	20	1.8×10^{-4}	0.71	Baumann, Lange (13)
DMF	0.24	25	2.21×10^{-4}	0.690	This work
DMF	0.46	30	1.2×10^{-4}	0.77	Shimura (8)
DMF	0.50	20	2.65×10^{-4}	0.72	Lange, Baumann (12)
DMF	0.50	25	2.51×10^{-4}	0.690	This work
THF	0	25	1.60×10^{-4}	0.706	Provder, Rosen (14)
THF	0	25	1.41×10^{-4}	0.700	Benoit (15)
THF	0	25	1.16×10^{-4}	0.73	Spychaj, et al (16)
THF	0	25	2.03×10^{-4}	0.681	This work

TABLE III. Molecular Structure – [η] Data for S/MA and S/MA/MM Samples

Sample	w_{MA}	w_{MM}	M_w $\times 10^{-3}$	[η] (dl/g) MEK	$^*M_w \times P_t$	$^*M_w \times P_c$
S/MA – 1 L	0.209	0	79.8	0.282	8.08×10^4	1.32×10^6
S/MA – 2 L	0.219	0	112	0.374	1.13×10^5	1.85×10^6
S/MA – 3 L	0.232	0	151	0.424	1.53×10^5	2.50×10^6
S/MA – 4 L	0.214	0	151	0.456	1.53×10^5	2.50×10^6
S/MA – 5 L	0.214	0	172	0.463	1.74×10^5	2.85×10^6
S/MA – 6 B	0.238	0	242	0.584	2.45×10^5	4.01×10^6
S/MA/MM – 7 L	0.173	0.08	170	0.43	1.72×10^5	2.81×10^6
S/MA/MM – 8 B	0.219	0.08	383	0.715	3.89×10^5	6.34×10^6
S/MA/MM – 9 L	0.273	0.07	222	0.55	2.26×10^5	3.69×10^6
S/MA/MM – 10 L	0.304	0.07	195	0.58	1.99×10^5	3.24×10^6
S/MA/MM – 11 B	0.268	0.07	155	0.46	1.58×10^5	2.57×10^6
S/MA/MM – 12 L	0.195	0.13	268	0.65	2.73×10^5	4.43×10^6

$^*P_t = (6.12 \times 10^{-2} \, w_{MA} + 4.00 \times 10^{-2} \, w_{MM} + 1)$

$^*P_c = (w_{MA} + 16.33)$

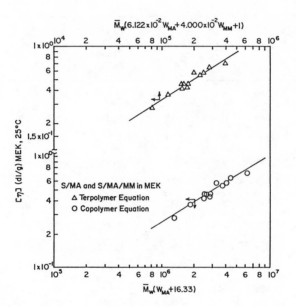

Figure 4.   Intrinsic viscosities of S/MA and S/MA/MM samples in MEK solvent vs. the terpolymer parameter, $M_w(6.122 \times 10^{-2} w_{MA} + 4.000 \times 10^{-2} w_{MM} + 1)$, and vs. the copolymer parameter, $M_w(w_{MA} + 16.33)$.

$$[\eta]_{25°C,MEK} = 5.67 \times 10^{-5} (w_{MA} + 16.33)^{0.608} M_w^{0.608} \quad (15)$$

with $R^2 = 0.973$, $\beta = 0.608$, and $J = 3.44 \times 10^{-3}$, in complete agreement with the terpolymer treatment. Reducing either eq.(14) or (15) to the polystyrene case (not necessarily justified by the composition range covered) gives $K = 3.1 \times 10^{-4}$ and $\alpha = 0.608$. Comparison of this result with the similar reduction of the SAN result to zero percent AN and with literature results, all given in Table II, shows quite good agreement. Thus, where such comparison is possible, the data demonstrate internal consistency between the polymer systems and verify the general copolymer and terpolymer equations, at least for these polymer-solvent systems.

Summary

A theoretical framework has been developed for expressing the intrinsic viscosity of a multispecies polymer formed by addition across single double bonds in the monomer species in terms of molecular weight and polymer composition. Based on the assumption that intrinsic viscosity is related to the number of carbon atoms in the polymer chain backbone, the general expression for the multispecies polymer of n monomer species has been derived and the specific cases of copolymers and terpolymers developed. Experimentally determined data for SAN in three solvents, DMF, THF, and MEK and for S/MA/MM in MEK were obtained and treated in the context of this theory, confirming its applicability, as well as defining the polymer composition ranges over which the equations may be validly applied for each solvent. Demonstration of general applicability to wide ranging copolymer and higher multispecies polymer types clearly requires testing of numerous other systems beyond those studied here. Moreover, the development of the theoretical formalism in no way obviates the requirement for experimental (empirical) determination for each multispecies polymer type-solvent combination of the values of the constants, involving interactions, which appear in the equations and of their range of applicability. It does, however, provide a rational framework for accounting for the polymer composition, as well as molecular weight, in the definition of the M-H-S equation, and it has special utility for SEC "universal calibration".

Acknowledgments

The author wishes to thank Drs. R. L. Kruse and A. R. Padwa for samples, as well as for useful discussions. The use of SEC molecular data originally obtained by Mr. R. Martin and Dr. A. S. Kenyon is acknowledged with thanks, as is the experimental assistance of Miss S. E. Vollrath and Mr. J. P. Chlastawa. Finally, the author wishes to thank the Monsanto Chemical Company for permission to publish this work.

Literature Cited

1. Yau, W. W.; Kirkland, J. J.; Bly, D. D. "Modern Size-Exclusion Liquid Chromatography", John Wiley & Sons, New York, 1979, pp.291-294.
2a. Benoit, H.; Grubisic, Z.; Rempp, P.; Decker, D.; Zilliox, J. G. J. Chim. Phys., 1966, 63, 1507.
2b. Grubisic, Z.; Rempp, P.; Benoit, H. J. Polym. Sci. B, 1967, 5, 753.
2c. Boni, K. A.; Sliemers, F. A.; Stickney, P. B. J. Polym. Sci. A2, 1967, 5, 221.
3. Kruse, R. L.; Padwa, A. R. J. Polym. Sci., Phys. Ed., 1983, 21, 1251
4. Ouano, A. C.; Kaye, W. J. Polym. Sci., A-1, 1974, 12, 1151.
5. McConnell, M. L. American Laboratory, May, 1978.
6. Martin, R., private communication.
7. Kenyon, A. S., private communication.
8. Shimura, Y. J. Polym. Sci., A-2, 1966, 4, 423.
9. Outer, P.; Carr, C. I.; Zimm, B. H. J. Chem. Phys., 1950, 18, 830.
10. Oth, J.; Desreux, V. Bull. soc. chim. Belges, 1954, 63, 285.
11. Bawn, C. E. H.; Freeman, C.; Kamaliddin, A. Trans. Farad. Soc., 1950, 46, 1107.
12. Lange, H.; Baumann, H. Angew. Makromol. Chemie, 1970, 14, 25.
13. Baumann, H.; Lange, H. Angew. Makromol. Chemie, 1969, 9, 16.
14. Provder, T.; Rosen, E. M. Separation Science, 1970, 5, 437.
15. Benoit, H. J. Chem. Phys., 1966, 63, 1507.
16. Spychaj, T.; Lath, D.; Berek, D. Polymer, 1979, 20, 437.
17. Goldwasser, J. M.; Rudin, A. J. Liq. Chromatog., 1983, 6, 2433-63.
18. Grinshpun, V.; Rudin, A. Proc. ACS Div. Polym. Matl.: Sci. & Eng., 1986, 54, 174-9.

RECEIVED March 10, 1987

Chapter 16

Dependence of Spreading Factor on the Retention Volume of Size Exclusion Chromatography

Rong-Shi Cheng, Zhi-Liu Wang, and Yang Zhao

Department of Chemistry, Nanjing University, Nanjing, People's Republic of China

The variance of the instrumental spreading function, i.e. the spreading factor of monodispersed polymer in a SEC column was determined experimentally with narrow MWD polystyrene standard samples by the method of simultaneous calibration. The dependence of the spreading factor on the retention volume deduced from a simple theoretical approach may be expressed by a formula with four physically meaningful and experimentally determinable parameters. The formula fits the experimental data quite well and the conditions for the appearance of a maximum spreading factor are explicable.

In the previous paper (1) a method for simultaneous calibration of molecular weight separation and instrumental spreading of SEC with characterized polymer standards was proposed. A thorough knowledge about how the spreading factor varies with the retention volume is of decided importance when applying broadening corrections for MWD determination. It is also useful for studying the pore surface structure of the SEC packings (2) and for deciding whether a resolvable peak of the chromatogram corresponds to a monodisperse species or not. In the present article a theoretical formula relating the spreading factor to the retention volume with four physically meaningful and experimentally determinable parameters is given and tested with experimental results.

THEORY

The variance of the spreading function, i.e. the spreading factor of monodisperse polymer σ_0^2 in a SEC column may be written as

0097–6156/87/0352–0281$06.00/0
© 1987 American Chemical Society

$$\sigma_0^2 = \sigma_{EX}^2 + \sigma_{SEC}^2 \qquad (1)$$

The first term σ_{EX}^2 is the combined contribution of the longitudinal dispersion and extracolumn effect to the spreading factor and may be regarded as a constant nearly independent on the molecular weight of polymer. The second term σ_{SEC}^2 is the contribution of the SEC process. According to the rate theory (3,4), σ_{SEC}^2 is proportional to the distribution coefficient K_{SEC} and inversely proportional to the diffusion coefficient D of the polymer in the pore

$$\sigma_{SEC}^2 \propto K_{SEC} / D \qquad (2)$$

The distribution coefficient K_{SEC} is defined as

$$K_{SEC} = (V_R - V_0) / V_i \qquad (3)$$

and the diffusion coefficient is empirically related to the molecular weight of the polymer as

$$D \propto M^{-\varepsilon} \qquad (4)$$

In Eq. 3 and 4, V_0 and V_i are the interstitial volume and total pore volume of the packings in the column respectively and ε is a constant nearly equal to one. The molecular weight is related to the retention volume by a linear calibration function $M(V_R)$:

$$\ln M = A_M - B_M V_R \qquad (5)$$

Substituting Eq. 3, 4 and 5 into Eq. 2 we have

$$\sigma_{SEC}^2 = c [(V_R - V_0)/V_i] \, Exp[\varepsilon (A_M - B_M V_R)] \qquad (6)$$

in which c is a proportionality constant. Putting

$$a = (c / V_i) \, Exp(\varepsilon A_M) \qquad (7)$$

$$b = \varepsilon B_M \qquad (8)$$

and substituting Eq. 6 into Eq. 1, we get

$$\sigma_0^2 = \sigma_{EX}^2 + a(V_R - V_0) \, Exp(-bV_R) \qquad (9)$$

which relates the spreading factor σ_0^2 to the retention volume V_R with four physically meaningful parameters σ_{EX}^2, V_0, a and b. The former two parameters σ_{EX}^2 and V_0 could be determined by separate experiments, for example by injecting a sample with high molecular weight exceeding the exclusion limit, or by the method of Groh and Halász (5). Parameter a and b may be estimated from calibration function and (or) from experimental data by curve fitting technique.

Eq. 9 predicts that a maximum spreading factor exists at a certain particular retention volume. Differentiating Eq. 9 with respect to V_R, we have

$$V_{R,max} = V_0 + 1/b \qquad (10)$$

$$\sigma^2_{0,max} = \sigma^2_{EX} + (a/b) \, Exp\{-(1+bV_0)\} \qquad (11)$$

Eq. 10 shows that $V_{R,max}$ depends on the value of V_0 and b. If both the interstitial volume and the slope of the calibration line are high, the maximum may escapes from one's observtion.

Experimental

Two series of narrow MWD polystyrene standards were used to calibrate a SN-01A GPC equipment, one of which (TSK) was supplied by Toyo Soda Co. and the other (NPS) was prepared and characterized in this laboratory. The column (3×1 M) were packed with NDG porous silica beads. Tetrahydrofurane was used as eluent. The elution volume was counted by a 2.60 ml. siphone tube.

Results and Discussion

The mean elution volume $\bar{V}$ and total variance σ^2_T of the experimental chromatogram of polystyrene standards were first calculated. The results are listed in Table I. By the method of simultaneous calibration of molecular weight separation and column dispersion (1) , the coefficients of the effective relation between the molecular weight and elution volume

$$M^*(V): \qquad \ln M = A_M^* - B_M^* V \qquad (12)$$

were evaluated for each standard with the aid of iteration starting from the experimental chromatogram, molecular weight and inhomogeneity data. The calibration function of the column was evaluated by linear regression between the mean elution volume $\bar{V}$ and the crosspoint molecular weight $M(\bar{V})$ yielding

$$\ln M = 22.724 - 0.342 \, V_R \qquad (13)$$

From the theoretical relationship between the coefficients of the effective relation and tha calibration function

$$A_M^* = A_M - (1 - \varsigma) B_M \bar{V} \qquad (14)$$

$$B_M^* = \varsigma \, B_M \qquad (15)$$

the parameter ς for each standard was then calculated. Finally the spreading factor was obtained according to the definition of parameter ς :

$$\varsigma^2 = (\sigma^2_T - \langle \sigma^2_0 \rangle) / \sigma^2_T \qquad (16)$$

The average spreading factor $\langle \sigma^2_0 \rangle$ of narrow MWD polystyrene standards may be regarded as the spreading factor σ^2_0 of a monodisperse polymer for which $V_R = \bar{V}$. All the results thus obtained are listed in Table II.

The variation of the spreading factor with retention volume is shown in Fig. 1. The existence of a maximum

284   DETECTION AND DATA ANALYSIS IN SIZE EXCLUSION CHROMATOGRAPHY

Table I. The Molecular Weight and SEC Data
          of Polystyrene Standards

Polymer	$\langle M \rangle_w \times 10^{-4}$	$\langle M \rangle_w / \langle M \rangle_n$	$\bar{V}$	σ_T^2
TSK-2	1.73	1.02	37.80	0.67
TSK-10	9.89	1.02	32.96	0.80
TSK-20	18.4	1.07	31.44	1.51
TSK-40	42.7	1.05	28.49	1.43
TSK-80	79.1	1.01	26.50	0.70
TSK-128	130	1.05	25.50	0.95
NPS-2	0.563	1.05	42.05	0.91
NPS-3	1.20	1.04	39.52	0.86
NPS-4	2.78	1.03	36.89	0.80
NPS-5	5.00	1.02	35.05	0.77
NPS-6	12.0	1.06	32.40	1.28
NPS-7	15.4	1.07	31.63	1.39

Table II. The Coefficients of the Effective relations
           and the Spreading Factors of Polystyrene
           Standards

Polymer	$A_M{}^*$	$B_M{}^*$	ξ	σ_0^2
TSK-2	16.24	0.172	0.500	0.50
TSK-10	16.68	0.157	0.462	0.63
TSK-20	18.73	0.211	0.623	0.92
TSK-40	18.17	0.184	0.535	1.02
TSK-80	16.60	0.114	0.329	0.68
TSK-128	19.81	0.226	0.663	0.53
NPS-2	19.15	0.253	0.746	0.43
NPS-3	21.00	0.299	0.873	0.48
NPS-4	19.27	0.247	0.723	0.52
NPS-5	21.91	0.316	0.928	0.57
NPS-6	20.50	0.270	0.794	0.72
NPS-7	19.38	0.233	0.685	0.84

spreading factor is obvious and in accord with some
literature results. Rearranging Eq. 9 we have

$$\ln (\sigma_0^2 - \sigma_{EX}^2)/(V_R - V_0) = \ln a - b V_R \qquad (17)$$

the logarithmic term should decreases linearly with V_R .
Such a plot is shown in Fig. 2 in which the value of σ_{EX}^2
and V_0 are estimated from the chromatograms of totally
excluded samples as 0.37 and 25.3 respectively. From the
intercept and slope of the line in Fig. 2, the parameter
a and b are evaluated and equal to 1190 and 0.309 respectively. The calculated curve of $\sigma_0^2(V_R)$ with these evaluated parameters is drawn in Fig. 1 too. The coincidence
with the experimental data is quite well.

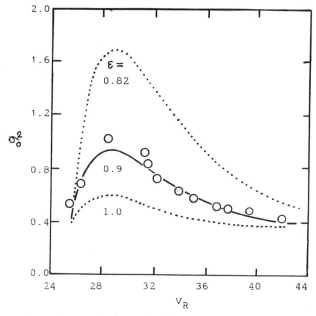

Figure 1.   Dependence of the spreading factor on the
the retention volume.

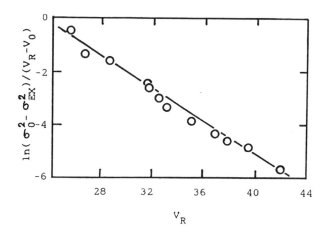

Figure 2.   Plot of $\ln(\sigma_0^2 - \sigma_{EX}^2)/(V_R - V_0)$ versus V_R .

It should be noticed that the value of parameter b
(0.309) is slightly smaller than the slope of the calibration function B_M (0.342) as expected by the theory.
From the relationship between b and B_M (Eq. 8), we get
ε = 0.90 for the present case. The calculated theoretical
$\sigma_0^2(V_R)$ curves are very sensitive to the value of the
parameter ε as shown in Fig. 1. It indicates that the
diffusion behavior of macromolecules in the pore of SEC
Packings can be studied in a quantitative way by systematic investigation of instrumental spreading effect.

Literature Cited

1. Rong-shi Cheng and Shu-qin Bo, ACS Symposium Series
   245, "SEC, Methodology and Characterization of Polymers
   and Related Materials", 1984; p. 125.
2. Rong-shi Cheng and Shu-qin Bo, Gaofenzi Tongxun
   (Polymer Communication, China), 1986 (5); 365.
3. M. Kubin, J. Chromatogr., 1975; 108, 1.
4. W. W. Yau, J. J. Kirkland and D. D. Bly, "Modern Size
   Exclusion Liquid Chromatography", 1979; p. 82.
5. R. Groh and I. Halász, Anal. Chem., 1981; 53, 1325.

RECEIVED May 15, 1987

Chapter 17

Correction for Instrumental Broadening in Size Exclusion Chromatography Using a Stochastic Matrix Approach

Based on Wiener Filtering Theory

L. M. Gugliotta, D. Alba, and G. R. Meira[1]

Intec (Conicet and Universidad Nacional del Litoral), (3000) Santa Fe, Argentina

The correction for non-uniform instrumental broadening in SEC is solved through a non-recursive matrix stochastic technique. To this effect, Tung's equation (1) must be reformulated in matrix form, and the measurements assumed contaminated with zero-mean noise. The proposed technique is based on an extension to time-varying systems of Wiener's optimal filtering method (1-3). The estimation of the corrected chromato gram is optimal in the sense of minimizing the estimation error variance. A test for verifying the results is proposed, which is based on a comparison between the "innovations" sequence and its corresponding expected standard deviation. The technique is tested on both synthetic and experimental examples, and compared with an available recursive algorithm based on the Kalman filter (4).

Most methods of correction for instrumental broadening in SEC (or hydrodynamic chromatography) are based on the deterministic integral equation due to Tung (5):

$$z(t) = \int_{-\infty}^{\infty} g(t,\tau)\, u(\tau)\, d\tau \qquad (1)$$

where t,τ: both represent elution time or elution volume;
      $z(t)$: is the base-line corrected chromatogram;
    $g(t,\tau)$: is the time-varying or non-uniform spreading function, which is built up by the set of unit mass impulse responses $g(t)$ of truly monodisperse polymers with different elution times τ; and
     $u(t)$: is the corrected chromatogram.
When $g(t,\tau)$ is considered time-invariant, then Equation 1 reduces to a convolution integral.

There are two basic problems associated to Equation 1:
i) the determination of the spreading $g(t,\tau)$; and

[1]Correspondence should be addressed to this author.

0097-6156/87/0352-0287$06.00/0
© 1987 American Chemical Society

ii) the estimation of $u(t)$, based on the knowledge of $z(t)$ and $g(t,\tau)$.

With respect to the spreading calibration, several methods have been suggested e.g. (6-14). Numerous techniques have been proposed for solving the inverse filtering problem represented by Equation 1, with different degrees of success e.g. (4,15-19). Only references (4), (18) and (19) make no assumptions on the shape of $g(t,\tau)$.

In this work, an inverse filtering technique based on Wiener's optimal theory (1-3) is presented. This approach is valid for time-varying systems, and is solved in the time domain in matrix form. Also, it is in many respects equivalent to the numerically "efficient" Kalman filtering approach described in (4). For this reason, a comparison between the two techniques will be made.

Theory

The Spreading Model. Considering the discrete version of Equation 1, and bearing in mind that all intervening functions are of finite length, then one may write:

$$z(k) = \sum_{k_0=-r}^{k_0=s} g(k,k_0) u(k_0) \qquad (k = 0,1,2,\ldots, n) \qquad (2)$$

where k,k_0: are the discrete equivalents of t and τ, respectively;
     $-r,s$: are the lower and upper limits of the sum in Equation 2, with non-zero values of the indicated product.

Let $\underline{z}$ and $\underline{u}$ denote column vectors such that:

$$\underline{z} = [z(0), z(1), \ldots, z(n)]^T \qquad (3a)$$

$$\underline{u} = [u(0), u(1), \ldots, u(n)]^T \qquad (3b)$$

$\underline{z}$ has normally more non-zero elements than $\underline{u}$. Even though the theory can be modified to allow for this fact, we assume for simplicity that $\underline{u}$ has the same number of components as $\underline{z}$. Therefore, one can write Equation 2 in matrix form as follows:

$$\underline{z} = G \underline{u} \qquad (4a)$$

with

$$G = \begin{bmatrix} g(0,0) & g(0,1) & \cdots & g(0,n) \\ g(1,0) & g(1,1) & \cdots & g(1,n) \\ \cdot & \cdot & \cdot & \cdot \\ \cdot & \cdot & \cdot & \cdot \\ \cdot & \cdot & \cdot & \cdot \\ g(n,0) & g(n,1) & \cdots & g(n,n) \end{bmatrix} \qquad (4b)$$

Note the following:
- In general, the first and the last elements of $\underline{u}$ will be zero.
- Each column of G contains an impulse response, with the impulse applied at the element in the diagonal of G. (See Figure 1 for a 3-D representation of a typical G matrix.)

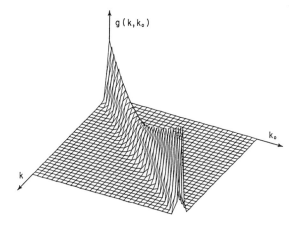

Figure 1: Tridimensional representation of a typical G matrix.

Hess and Kratz ($\underline{6}$) tried to estimate $\underline{u}$ directly from Equations 4 as follows:

$$\hat{\underline{u}} = G^{-1} \underline{z} \tag{5}$$

In general, this operation is numerically ill-conditioned, leading to incorrect results. (The degree of ill-conditioning may be measured by the condition number, defined by the ratio of the modulus of the largest to the smallest eigenvalue of G).

A stochastic version of Equation 4a may be written:

$$\underline{z} = G \underline{u} + \underline{v} \tag{6a}$$

or

$$\underline{z} = \underline{y} + \underline{v} \tag{6b}$$

where $\underline{v} = [v(0),..., v(n)]^T$: is a zero-mean additive noise;

$\underline{y}$: is the noise-free measured chromatogram; and

$\underline{z},\underline{u},\underline{y}$: will be assumed zero-mean stochastic variables.

In what follows, we shall seek a restoring matrix H such that the estimate $\underline{u}$ is calculated through:

$$\hat{\underline{u}} = H \underline{z} \tag{7}$$

Let $\underline{e}_u = (\underline{u}-\hat{\underline{u}})$ be the estimation error associated with $\underline{u}$. The estimate $\hat{\underline{u}}$ is chosen in such a way that the corresponding mean square error $\overline{E[(u-\hat{u})^T(u-\hat{u})]}$ is minimized.

The Input Estimation Through a Wiener Filtering Approach. Equation 6a represents a time-varying linear filter with a measurement noise, and the statistics of such noise may be considered non-stationary. Simply stated, the optimal inverse filtering problem is this: assuming that a signal is first distorted through a linear filter of known characteristics and then contaminated with an additive noise, what linear operation on the resulting measurement will yield the best estimation of the original signal?. "Best" in this case means minimum mean-square error. This branch of filtering began with N. Wiener's work in the 1940's ($\underline{1}$). R.E. Kalman then made an important contribution in the early 1960's; by providing an alternative approach to the same problem using state-space methods ($\underline{20-21}$).

N. Wiener's solution was originally derived in the frequency domain for time-invariant systems with stationary statistics. In what follows, a matrix solution derived from such approach but developed in the time domain for time-varying systems and non-stationary statistics will be presented ($\underline{22-23}$). An expression for the required transformation H in Equation 7 will be obtained. In all that follows, we shall denote with $\hat{\underline{u}}$ the best estimate of $\underline{u}$, i.e. an estimate such that:

$$E[(\underline{u}-\hat{\underline{u}})^T(\underline{u}-\hat{\underline{u}})] \leqslant E[(\underline{u}-\tilde{\underline{u}})^T(\underline{u}-\tilde{\underline{u}})] \tag{8}$$

where $\tilde{u}$ is any suboptimal estimate of u. The principle of orthogonality (24-26), states that Equation 8 will be verified if the estimation error vector is orthogonal to the measurements. In other words, the following must be true:

$$E[(\underline{u}-\hat{\underline{u}})\underline{z}^T] = 0 \qquad (9)$$

Substituting Equation 6a and Equation 7 into Equation 9 and operating, one obtains:

$$E[\underline{u}\underline{u}^T] \ G^T + E[\underline{u}\underline{v}^T] = H \ E[\underline{z}\underline{z}^T] \qquad (10)$$

We shall assume the input u uncorrelated with v, i.e.:

$$E[\underline{u}\underline{v}^T] = E[\underline{v}\underline{u}^T] = 0 \qquad (11)$$

Let Σ_u and Σ_z be the covariance matrices corresponding to u and z, respectively. (Such matrices are in general non-stationary). Thus, Equation 10 may be written:

$$\Sigma_u \ G^T = H \ \Sigma_z \qquad (12)$$

From:

$$\Sigma_z = E[(G\underline{u}+\underline{v})(G\underline{u}+\underline{v})^T] \qquad (13)$$

and bearing in mind Equation 11, one finds:

$$\Sigma_z = G \ \Sigma_u \ G^T + \Sigma_v \qquad (14)$$

Substituting Equation 14 into Equation 12 and operating, one finally arrives at:

$$H = \Sigma_u \ G^T \ [G \ \Sigma_u \ G^T + \Sigma_v]^{-1} \qquad (15)$$

In other words, the optimal estimate may be calculated through:

$$\hat{\underline{u}} = \Sigma_u \ G^T \ [G \ \Sigma_u \ G^T + \Sigma_v]^{-1} \ \underline{z} \qquad (16)$$

Note the following:
- For any arbitrary G, the existence of $[G \ \Sigma_u \ G^T + \Sigma_v]^{-1}$ is ensured by the invertibility of Σ_v.
- Adopting $\Sigma_u=qI$ and $\Sigma_v=0$, then Equation 16 reduces to Equation 5.
- With $\Sigma_u=qI$ and $\Sigma_v=rI$, Equation 16 has a format which is identical to the solution derived in (27) through a deterministic minimum least squares approach for time-invariant systems. This is to be expected, because the Wiener filtering technique may be in fact included as part of the general theory of least squares.

The Filter Adjustment. The computation of $\hat{\underline{u}}$ through Equation 16 involves the prespecification of Σ_u and Σ_v. These matrices are in general symmetric; and the simplification of considering both v and u white noises has been found to provide satisfactory results. Thus, Σ_u and Σ_v will be assumed diagonal.

The statistics of $\underline{v}$ may be considered stationary with sound physical basis. Therefore, we shall simply adopt:

$$\Sigma_v = r \, I \qquad (17)$$

where the scalar r may be obtained from the sample variance of the chromatogram baseline noise. Note that for any positive r, the invertibility of $[G \, \Sigma_u \, G^T + \Sigma_v]$ in Equation 16 is theoretically ensured.

Consider now the estimation of the diagonal elements of Σ_u. The following assumptions can be made:

a) Take the variance of u(k) to be constant. In this case, and remembering that u(k) is assumed of zero mean, one may write:

$$\Sigma_u = q \, I \qquad (18a)$$

where the value of q may be simply estimated from the measurement z(k) as follows:

$$q = \frac{1}{n} \sum_{k=0}^{n} [z(k)]^2 \qquad (18b)$$

b) Allow now the variance of u(k) to be time-varying. (This is more reallistic than before). Call:

$$\Sigma_u = \text{diag.} \, [q(0), \, q(1), \, \ldots \, q(n)] \qquad (19)$$

Here, we can estimate q(k) in several ways, for example:

$$q(k) = C_1 \, [z(k)]^2 \qquad (20)$$

or

$$q(k) = C_2 \, [\tilde{u}(k)]^2 \qquad (21)$$

where C_1, C_2 are positive constants, and $\tilde{u}(k)$ is any other suboptimal estimation of $\underline{u}$.

The Solution Validation. Obvious conditions that the resultant solution $\hat{\underline{u}}$ must satisfy are: a) $\hat{\underline{u}}$ must be non-negative; b) the operation $G\hat{\underline{u}}$ should provide a noise-free measured function; and c) the areas under the measured and the corrected chromatograms must be equal. It should be emphasized that condition b) is a necessary but not sufficient for good results. Apart from the mentioned checks, a validation procedure based on the analysis of the innovations will now be presented.

Consider first the covariance matrix Σ_{e_u} corresponding to the estimation error $\underline{e}_u$, i.e.:

$$\Sigma_{e_u} = E[(\underline{u}-\hat{\underline{u}})(\underline{u}-\hat{\underline{u}})^T] \qquad (22)$$

Substituting Equation 6a into Equation 7 and the latter in turn into Equation 22, one obtains:

$$\Sigma_{e_u} = \Sigma_u - 2 \, H \, G \, \Sigma_u + H \, G \, \Sigma_u \, G^T \, H^T + H \, \Sigma_v \, H^T \qquad (23)$$

The innovations sequence $\underline{e}_z$ is defined by:

$$\underline{e}_z = \underline{z} - \hat{\underline{z}} \tag{24}$$

and therefore,

$$\underline{e}_z = \underline{z} - G\,\hat{\underline{u}} \tag{25}$$

because the best estimate for $\underline{z}$ is $\hat{\underline{y}} = G\hat{\underline{u}}$, since $\underline{v}$ is zero mean. Substituting Equation 6a into Equation 25 yields:

$$\underline{e}_z = G\,\underline{e}_u + \underline{v} \tag{26}$$

The corresponding covariance matrix is found substituting Equation 23 into:

$$\Sigma_{e_z} = E[(G\,\underline{e}_u + \underline{v})\,(G\,\underline{e}_u + \underline{v})^T] \tag{27}$$

and the final result is:

$$\Sigma_{e_z} = G\,\Sigma_u\,G^T - 2\,G\,H\,G\,\Sigma_u\,G^T + G\,H\,G\,\Sigma_u\,G^T\,H^T\,G^T + G\,H\,\Sigma_v\,H^T\,G^T +$$
$$+ \Sigma_v \tag{28}$$

The proposed check consists in matching the innovations sequence obtained from Equation 25 with the corresponding expected time-varying variance provided by Equation 28. If the innovations sequence is assumed zero-mean Gaussian white, then $e_z(k)$ should be within the $\pm\sigma_{e_z}(k)$ bounds for approximately two thirds of the time. ($\sigma_{e_z}(k)$ represents the standard deviation of $e_z(k)$, found by square rooting the diagonal elements of Σ_{e_z}).

Note that the proposed check must be perfomed after having obtained the estimation of $\underline{u}$. In contrast, in the Kalman filter technique (4), the corresponding values of $\underline{e}_z$ and Σ_{e_z} may be recursively calculated along with the input estimate.

Examples of Application

In order to compare the present technique with the method based on the Kalman filter (4), the same examples presented in that publication will be attempted. The first two examples are synthetic, while the third is based on real experimental data. All examples were solved by means of a VAX 11/780 computer programmed in FORTRAN 77. Routines for matrix operation from the IMSL package (28) were utilized.

Example 1. By processing the curve $u(k)$ shown in Figure 2a through a time-varying filter defined by the set of impulse responses of Figure 1, a noise-free chromatogram $y(k)$ is obtained. This curve was then contaminated with Gaussian white noise of a relatively low variance (10^{-5}), to provide $z(k)$. Clearly, the best estimate for r is 10^{-5}, and in this case a constant value for $q = 5 \times 10^{-3}$ was adopted by trial and error, providing an acceptable compromise between the different checks.

Let $\hat{u}_K(k)$ be the optimal estimate obtained through the Kalman approach. $\hat{u}(k)$ and $\hat{u}_K(k)$ are also shown in Figure 2a. The innovations corresponding to $\hat{u}(k)$ are represented in Figure 2b. This example was solved assuming noisy baseline sections before and after the peak as part of the chromatogram. For this reason, and because q was assumed constant, oscillations are observed in $\hat{u}(k)$ and $\hat{u}_K(k)$ in those sections of the curve. In both techniques, better estimations are obtained if q is adopted time-varying through Equation 20. In this case, the mentioned oscillations around the baseline sections before and after the peak disappear.

Example 2. This example was first suggested by Chang and Huang (29), and attemped later on by Hamielec and co-workers (19). The problem is illustrated by Figure 3, which represents the following: u(k), the uniform spreading function g(k), the broadened curve z(k), and the recuperated $\hat{u}_2(k)$ by method 2 proposed in (19). The solution shown in Figure 3 is practically coincident with that of (29), and with that of method 1 in (19). Clearly, these techniques are unable to appropriately recover the double-peaked input.

This problem was solved adopting the same values for r and q as in (4), i.e.: r=0.1 and q calculated through Equation 20 with C_1=1. The results are shown in Figure 4a, where the original u(k) is compared to the estimates obtained through the proposed technique and through the Kalman approach (Figure 10a of (4)). Figure 4b illustrates the innovations test.

Example 3. Curve z(k) in Figure 5 represents the chromatogram of a PS standard of MW=525, when fractionated through an A-802 Shodex column mounted on a Series 3-B Perkin Elmer liquid chromatograph. The chromatogram of pure benzene g(k) is adopted as the uniform spreading function. The polymer sample is expected to be integrated by the first PS oligomers, with preponderance of the pentamer. Ideally, delta functions ought to be recuperated, with the highest peak at a molecular weight of 520.

Here, a value of $r=5\times10^{-5}$ was adopted, and q was calculated through Equation 20 with C_1=0.75. In Figure 5, the result of the present technique is compared to the result in Figure 12a of (4). As with all previous examples, the estimated noise-free chromatogram y(k) is practically coincident with the measured z(k).

Conclusions

The proposed technique is numerically "robust", and its results are comparable to those obtained through a recursive method based on the Kalman filter (4). It should be noted that because the present technique utilizes all of the information simultaneously, the results have been compared to those of the optimal smoother estimates in (4), which are "better" than the true filtered estimates.

The main advantage of the stochastic matrix approach is the simplicity for its computer implementation. Equation 17 directly provides the desired result, and Equation 28 is the basis of a validation test which may or may not be performed according to previous experience. In other words, the proposed method is conceptually and practically easier to implement than the Kalman counterpart. The

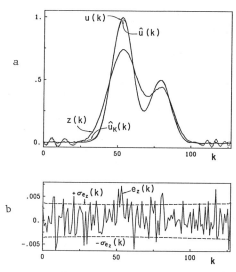

Figure 2: Example 1: a) Comparison between the "true" input $u(k)$, the estimation of that input through the present technique $\hat{u}(k)$ and the same estimation through the method described in (4) $\hat{u}_K(k)$; b) Innovations sequence and $\pm\sigma_{e_z}(k)$ bounds corresponding to $\hat{u}(k)$.

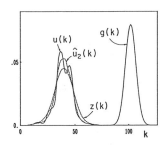

Figure 3: Example 2: (after Hamielec and co-workers (19)).

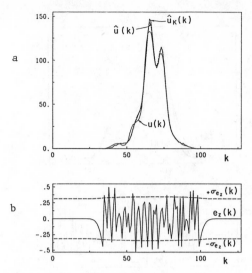

Figure 4: Example 2: a) Comparison of present results with those in (4); b) Validation test for û(k).

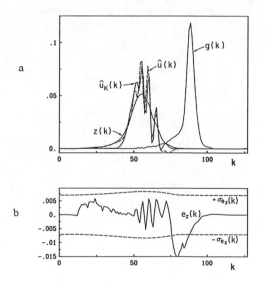

Figure 5: Example 3: a) Experimental chromatogram, spreading function and comparison of present results with those in (4); b) Validation test for û(k).

principal drawback of the present technique is its relatively high computational cost, both in memory and computation time. Typically, in order to solve a chromatogram of 128 points with a time-varying $q(k)$, a computation time of 5 mins. was required to estimate $u(k)$, and 4.5 more mins. were necessary for the validation test.

A point that has not been investigated is the possibility of considering $u(k)$ a coloured noise instead of white noise, and therefore a non diagonal Σ_u. For example, the choice of a tridiagonal Σ_u would imply the assumption of $u(k)$ a random walk process. On the one hand, by imposing a correlation among successive values of $u(k)$, the flexibility of the output is reduced, and for example a delta function could not be recuperated. On the other hand, smoother outputs and better solutions could be obtained if good "a priori" estimations of the real autocorrelations of $u(k)$ could be provided.

Finally, it should be noted that apart from its use in chromatographic data treatment, inverse filtering techniques such as that described in this work have also potential applications in other areas of polymerization engineering, (see for example (30) and (31)).

Acknowledgments

We would like to thank Mr. M. Brandolini for his help with the experimental work, CONICET and U.N.L. for their financial support and Dr. J.F. Weisz for revising the manuscript.

Literature Cited

1. Wiener, N., "Extrapolation, Interpolation and Smoothing of Stationary time Series"; J. Wiley and Sons, Inc.: New York, 1949, p. 163.
2. Helstrom, C.W., J. Opt. Soc. Am., 1967, 57, 297.
3. Sondhi, M.M., Proc. IEEE, 1972, 60, 842.
4. Alba, D. and Meira, G.R., J. Liq. Chromatogr., 1984, 7(14), 2833.
5. Tung, L.H., J. Appl. Polym. Sci., 1966, 10, 375.
6. Hess, M. and Kratz, R.F., J. Polym. Sci., Part A-2, 1966, 4, 731.
7. Husain, A., Hamielec, A.E. and Vlachopoulos, J., J. Liq. Chromatogr., 1981, 4, 459.
8. Tung, L.H., Moore, J.C. and Knight, G.W., J. Appl. Polym. Sci., Part A-2, 1966, 10, 1261.
9. Tung, L.H. and Runyon, J.R., J. Appl. Polym. Sci., 1969, 13, 2397.
10. Waters, J.L., J. Polym. Sci., Part A-2, 1970, 8, 411.
11. Grubisic-Gallot, Z., Marais, L. and Benoit, H., J. Polym. Sci., Polym. Physics Edition, 1976, 14, 959.
12. Gruneberg, J. and Klein, J., J. Liq. Chromatogr., 1980, 3, 1593.
13. McCrakin, F.L. and Wagner, H.L., Macromolecules, 1980, 13, 685.
14. Alba, D. and Meira, G.R., J. Liq. Chromatogr., (in press).
15. Vozka, S. and Kubin, M., J. Chromatogr., 1977, 139, 225.
16. Hamielec, A.E., J. Liq. Chromatogr., 1980, 3(3), 381.
17. Hamielec, A.E., Ederer, H.J. and Ebert, K.H., J. Liq. Chromatogr., 1981, 4(10), 1697.
18. Chang, K.S. and Huang, R.Y.M., J. Appl. Polym. Sci., 1972, 16, 329.
19. Ishige, T., Lee, S.I. and Hamielec, A.E., J. Appl. Polym. Sci., 1971, 15, 1607.

20. Kalman, R.E., Trans. ASME, Series D, J. Basic Eng., 1960, 82, 35.
21. Kalman, R.E. and Bucy, R.S., Trans. ASME, Series D, J. Basic Eng., 1961, 83, 95.
22. Booton, R.C., Proc. IRE, 1952, 40, 977.
23. Davis, M.C., IEEE Trans. on Automatic Control, 1963, AC-8, 196.
24. Papoulis, A., "Probability, Random Variables and Stochastic Processes", McGraw-Hill 1965.
25. Srinath, M.D., Rajasekaran, P.K.; "An Introduction to Statistical Signal Processing with Applications", Wiley 1979.
26. Anderson, B.D.O. and Moore, J.B., "Optimal Filtering", Prentice Hall 1979.
27. Rosen, E.M. and Provder, T., J. Appl. Polym. Sci., 1971, 15, 1687.
28. The International Mathematical Statistical Libraries, Inc. 1980.
29. Chang, K.S. and Huang, R.Y.M., J. Appl. Polym. Sci., 1969, 13, 1459.
30. Couso, D., Alassia, L. and Meira, G.R., J. Appl. Polym. Sci., 1985, 30(8), 3249.
31. Gugliotta, L.M. and Meira, G.R., Die Makromoleculare Chemie, (in press).

RECEIVED February 26, 1987

INDEXES

Author Index

Affiliation Index

Subject Index

300

Production by Barbara J. Libengood
Indexing By Deborah H. Steiner
Jacket Design by Carla L. Clemens

Elements typeset by Hot Type Ltd., Washington, DC
Printed and bound by Maple Press Co., York, PA

Recent Books

Personal Computers for Scientists: A Byte at a Time
By Glenn I. Ouchi
276 pp; clothbound; ISBN 0–8412–1000–4

The ACS Style Guide: A Manual for Authors and Editors
Edited by Janet S. Dodd
264 pp; clothbound; ISBN 0–8412–0917–0

Silent Spring Revisited
Edited by Gino J. Marco, Robert M. Hollingworth, and William Durham
214 pp; clothbound; ISBN 0–8412–0980–4

Chemical Demonstrations: A Sourcebook for Teachers
By Lee R. Summerlin and James L. Ealy, Jr.
192 pp; spiral bound; ISBN 0–8412–0923–5

Phosphorus Chemistry in Everyday Living, Second Edition
By Arthur D. F. Toy and Edward N. Walsh
362 pp; clothbound; ISBN 0–8412–1002–0

Pharmacokinetics: Processes and Mathematics
By Peter G. Welling
ACS Monograph 185; 290 pp; ISBN 0–8412–0967–7

Liquid Membranes: Theory and Applications
Edited by Richard D. Noble and J. Douglas Way
ACS Symposium Series 347; 196 pp; ISBN 0–8412–1407–7

*Design Considerations for Toxic Chemical
and Explosives Facilities*
Edited by Ralph A. Scott, Jr. and Laurence J. Doemeny
ACS Symposium Series 345; 318 pp; ISBN 0–8412–1405–0

*Metal Complexes in Fossil Fuels: Geochemistry,
Characterization, and Processing*
Edited by Royston H. Filby and Jan F. Branthaver
ACS Symposium Series 344; 436 pp; ISBN 0–8412–1404–2

Sources and Fates of Aquatic Pollutants
Edited by Ronald A. Hites and S. J. Eisenreich
Advances in Chemistry Series 216; 558 pp; ISBN 0–8412–0983–9

Nucleophilicity
Edited by J. Milton Harris and Samuel P. McManus
Advances in Chemistry Series 215; 494 pp; ISBN 0–8412–0952–9

For further information and a free catalog of ACS books, contact:
American Chemical Society
Distribution Office, Department 225
1155 16th Street, NW, Washington, DC 20036
Telephone 800-227-5558